More on Counting

Xing Zhou

http://www.mathallstar.org

Copyright © 2015 by Xing Zhou. All rights reserved.

No part of this book may be reproduced, distributed or transmitted in any form or by any means, including photocopying, scanning, or other electronic or mechanical methods, without written permission of the author.

To promote education and knowledge sharing, reuse of some contents of this book may be permitted, courtesy of the author, provided that: (1) the use is reasonable; (2) the source is properly quoted; (3) the user bears all responsibility, damage and consequence of such use. The author hereby explicitly disclaims any responsibility and liability; (4) the author is notified in advance; and (5) the author encourages, but does not enforce, the user to adopt similar policies towards any derived work based on such use.

Please visit the website `http://www.mathallstar.org` for more information or email `contact@mathallstar.org` for suggestions, comments, questions and all copyright related issues.

use your mobile device to scan this QR code for more resources including books, practice problems, online courses, and blog.

This book was produced using the LaTeX system.

Contents

1	**Introduction**	**1**
	1.1 Contents	1
	1.2 Notation	2
2	**Bijection**	**3**
	2.1 Bijection Explained	3
	2.2 Examples	4
	2.3 Practice	8
3	**Recursion**	**15**
	3.1 Introduction	15
	3.2 More Examples	16
	3.3 Practice	21
4	**Integer Solution and Balls-in-Boxes**	**25**
	4.1 The Integer Solution Model	25
	4.2 Count Integer Solutions	27
	4.3 Other Situations	32
	4.4 The Balls-in-Boxes Model	34
	4.5 Practice	36
5	**Combinatorial Identities**	**41**
	5.1 Binomial / Multinomial Theorem	41
	5.2 The Counting Method	42
	5.3 The Coefficient Method	45
	5.4 Pascal Triangle	46
	5.5 Modular Properties	48
	5.6 The Special Value Method	49
	5.7 Hockey Stick Identity	53
	5.8 Change the Index	56
	5.9 Generalized Binomial Theorem	58
	5.10 Practice	61
6	**Generating Function**	**65**

CONTENTS

- 6.1 Introducing Generating Function 65
- 6.2 Generating Function Properties 67
- 6.3 Useful Conclusions and Techniques 68
- 6.4 Integer Solution Generalization 73
- 6.5 More Examples 75
- 6.6 Practice 77

Appendices   81

A Solutions   83
- A.1 *Chapter 1* 84
- A.2 *Chapter 2* 85
- A.3 *Chapter 3* 101
- A.4 *Chapter 4* 111
- A.5 *Chapter 5* 121
- A.6 *Chapter 6* 128

Preface

Welcome to Math All Star© series!

Math All Star originates from a series of lectures given to a group of gifted middle school students with a love for mathematics and an interest in participating in competitions such as MathCounts, AMC, and AIME. These lectures aim to strengthen their problem-solving abilities and to introduce effective techniques that are not typically taught in the classroom.

As the popularity of Math All Star grew, the author began to upload lecture materials to create online courses, thereby providing students with the opportunity to progress at their own paces.

Since then, course materials have constantly been reviewed and updated to reflect student feedback and the observations made during lectures. Recent competition problems are also continuously analyzed and referenced to ensure the relevance of the contents. These course materials are the foundations of this Math All Star series.

Because competition math is a diversified subject that covers both a wide breadth and depth of topics, it is quite challenging to effectively cover all the material in one book that is appropriate for every interested student. Consequently, the author has decided to write a series of books, with each one focusing on a particular topic. Students are encouraged to pick and choose where to begin, depending on their individual skill levels and needs.

CONTENTS

In addition to these books, the Math All Star website provides extra practice problems and serves as a highly recommended supplemental learning resource.

If there are any questions, comments, or concerns, please visit the website or email `contact@mathallstar.org`.

Happy learning!

To visit the Math All Star website, scan this QR code or go directly to
`http://www.mathallstar.org`

Chapter 1

Introduction

1.1 Contents

This book is not the 2nd edition of the first *Counting* book. Instead, it contains topics which are built upon the contents discussed in the first book. Generally speaking, this book targets AIME, national and international level competitions. Some techniques are quite advanced.

Bijection is a generalization of the symmetry method which is discussed in the first book. While symmetry usually refers to two symmetric parts of the same nature, bijection utilizes one to one mapping among any two sets even if they appear to be completely unrelated. In fact, solutions to some challenging problems discussed in the first book, such as those in the modeling section, essentially are applications of bijection.

The recursion method solves a counting problem by first establishing a recursive relation and then solving this recursion to get the final answer. For example, the Hanoi Tower is a classic problems which can be best tackled using the recursion method. To solve a recursive relation is usually the same as solving a linear regression sequence. The latter is discussed in the book Competition Algebra.

Problems which are related to the integer solution pattern have appeared more and more frequently in recent years. Though necessary solving techniques are covered in the first book, this pattern deserves a more thorough discussion. Counting integer solutions also serves as an introduction to the more powerful generating function method.

Combinatorial identity is an important building block to solve some challenging counting problems. Some of these identities are well known basic relations, and others are not. It is difficult and unnecessary to remember all of these identities. However, it is essential to master core relations and corresponding problem solving techniques such as the special value method, coefficient method, etc.

Generating function will be discussed at the end. It is an advanced general purpose technique that can solve a wide range of math problems, including counting related ones. Usually, generating functions are not required in entry level competitions. However, it is beneficial for those students who aim to excel in national or even international level math contests.

1.2 Notation

In this book, $\binom{n}{k}$ is used instead of C_n^k. These two notations are interchangeable. However, the former style is more convenient when the multinomial theorem is used.

Chapter 2

Bijection

2.1 Bijection Explained

Bijection is a one-to-one mapping between elements in two sets $\mathbb{A}$ and $\mathbb{B}$. Every element in $\mathbb{A}$ has one and only one corresponding element in $\mathbb{B}$, and vice versa. If such a mapping relationship is one way only, then it is called either injection or surjection depending on the mapping direction.

An example of bijective function is permutation which maps a collection of elements to itself by just changing the location indexes of the entries.

Two sets having bijective relation are equivalent from counting perspective because the numbers of elements in these two sets are the same. Therefore, bijection can be a useful technique when a difficult to count set can be transformed to an easier to count equivalent set.

The symmetry technique discussed in the first *Counting* book is one type of bijections. Typically, the symmetry method identifies a one-to-one mapping between two disjoint subsets of the original set

Chapter 2: Bijection

which is to be counted. These two subsets are mutually exclusive and their union equals the original set. Therefore, the total count exactly doubles that of either subset. If one of these two subsets is easy to count, then the total count of the original set is simply twice as many.

Bijection goes further by exploring a one-to-one mapping between two different sets. If such a bijective relation can be established and the mapped set is easy to count, then the count of the original set can be obtained indirectly via this way. For instance, the solution to the following example from the first *Counting* book essentially employs the bijection method.

Two schools hold an annual chess tournament. Each team has 10 participants with a pre-determined order to play. In each round, the loser is eliminated and the winner plays against the next player from the opponent team. When all the 10 players of one team are eliminated, the contest finishes.

If we record the progress of the entire tournament (i.e. who plays against whom, and in which round), how many different outcomes are possible?

While symmetric relationship within the same set is often intuitive, identifying an appropriate bijective set can be hard and requires experience. The next section will present a couple of examples.

2.2 Examples

The key to use the bijection method is to find an easier-to-count equivalent set.

Chapter 2: Bijection

Example 2.2.1

Let n be a positive integer. Find the number of paths from $(0,0)$ to (n,n) on an $(n \times n)$ grid using only unit up and right steps so that the path stays in the region $x \geq y$.

(ref: 4355)

Counting the number of shortest paths in a rectangular $(m \times n)$ grid is a basic pattern. The answer is $\binom{m+n}{n}$ or $\binom{m+n}{m}$ because the problem is equivalent to selecting m unit right steps, or n unit up steps, from totally $(m+n)$ steps. If there are some blockades on the grid, then the number of good paths can be obtained using the lattice method. These two topics are covered in the first book.

In *Example 2.2.1*, the catch is the restriction on the valid region. Given the condition as $x \geq y$, it is natural to consider using the bijection method. Note that the answer to this problem is not half of the total number of unrestricted paths.

Solution

Let's first count the number of bad paths, i.e. those paths which go beyond the line $y = x$. Assuming P is such a bad path. Given P goes beyond the line $y = x$, it must touch one or more points of the line $y = x + 1$. Let the first such point be M. Then construct a mirroring path from $(0,0)$ to M with respect to the line $y = x + 1$ and keep the remaining unchanged. Name the new path as P'. This is shown in the diagram below.

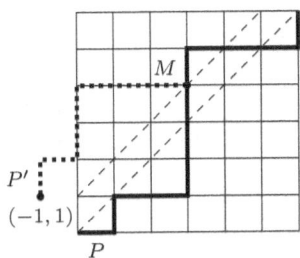

Chapter 2: Bijection

The mirroring point of $(0,0)$ is $(-1,1)$, so P' is a path from $(-1,1)$ to (n,n). Every bad path P will have a unique mirroring path P'. Conversely, every path P' from $(-1,1)$ to (n,n) will have a unique mirroring path P from $(0,0)$ to (n,n). Because P and P' share a point M on $y=x+1$, thus P must go outside valid region.

Because the number of path P' is $\binom{(n+1)+(n-1)}{n-1}$, therefore there are this many bad paths P.

Meanwhile, there are totally $\binom{n+n}{n}$ paths from $(0,0)$ to (n,n). It follows that the number of good paths equals

$$\binom{2n}{n} - \binom{2n}{n-1} = \boxed{\frac{1}{n+1}\binom{2n}{n}}$$

Done.

The value of

$$\frac{1}{n+1}\binom{2n}{n} = \frac{(2n)!}{n!(n+1)!} = \prod_{k=2}^{n} \frac{n+k}{k}$$

is called *Catalan number*. It is an interesting number because many counting problems are related to this number.

The next example is a good illustration of the bijection method since its solution is based on establishing a one-to-one mapping with an already solved problem.

Example 2.2.2

Find the number of expressions containing n pairs of parentheses which are correctly matched. For example, when $n=3$, the answer is 5 because all the legimate expressions are

$$((())),\ (()()),\ (())(),\ ()(()),\ ()()()$$

(ref: 4356)

Chapter 2: Bijection

Solution

A bijection can be established between this problem and the one in *Example 2.2.1*. (Constructing such a mapping will be explained below.) Consequently, the answer to this problem is

$$\boxed{\frac{1}{n+1}\binom{2n}{n}}$$

On an $(n \times n)$ grid, we can interpret the unit right step as the left parenthesis and the up step as the right parenthesis. Because it is a $(n \times n)$ grid, a path will have exactly n pairs of parentheses. Meanwhile, these parentheses are correctly matched if and only if at any point there are no more right parentheses than the left ones, which is equivalent to saying no more up steps than the right steps. This is the same as restricting the path to be in the region constrained by the condition $x \geq y$.

Done.

2.3 Practice

Practice 1

Explain the identity $\binom{n}{k} = \binom{n}{n-k}$ using bijection.

Practice 2

Given a convex n-polygon, what is the max number of intersection points its diagonals can form? (Vertices do not count.)

(ref: 4364)

Practice 3

How many fractions in simplest form are there between 0 and 1 such that the products of their denominators and numerators equal 20!?

(ref: 2717)

Practice 4

A child builds towers using identically shaped cubes of different colors. How many different towers with a height of 8 cubes can the child build with 2 red cubes, 3 blue cubes, and 4 green cubes? (One cube will be left out.)

(ref: 4385 - AMC10)

Chapter 2: Bijection

Practice 5

There are $n \geq 6$ points on a circle, every two points are connected by a line segment. No three diagonals are concurrent. How many triangles are created by these sides and diagonals?

(ref: 4368 - China)

Practice 6

John walks from point A to C while Mary goes from point B to D. Both of them will move along the grid, either right or up, so they take shortest routes. How many different possibilities are there such that their routes do not intersect?

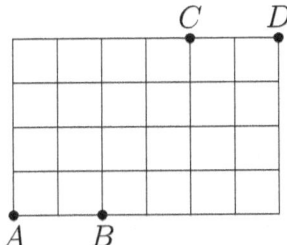

(ref: 3159)

Practice 7

Let N be the number of non-decrease sequences of length n and each element is a non-negative integer not exceeding d. Show that N equals the number of non-negative solutions to the following equation:

$$x_0 + x_1 + \cdots + x_d = n$$

Chapter 2: Bijection

Practice 8

A partition of a positive integer n is to write n as a sum of some positive integers. Let k be a positive integer. Show that the number of partitions of n with exactly k parts equals the number of partitions of n whose largest part is exactly k.

(ref: 4352)

Practice 9

Show that the number of partitions of a positive integer n into distinct parts is equal to the number of partitions of n where all parts are odd integers.

(ref: 4353)

Practice 10

Let positive integers n and k satisfy $n \geq 2k$. How many k-sided convex polygons are there whose vertices are those of an n-sided convex polygon and edges are diagonals of the same n-polygon.

(ref: 4362)

Practice 11

A tree is a structure which grows from a single root node. By convention, the root node is usually placed at the top. Then, new nodes, referred as children nodes, can be attached to an existing node, referred as parent node, by edges. The root node has no parent and all other nodes have exactly one parent. A node may have any number of children nodes, including no child at all. No looping chain of nodes is permitted in this structure. For example, there are exactly 5 different types of trees with 4 nodes. They are shown below. The goal is to find the number of different types of tree with n nodes where n is a positive integer greater than 1.

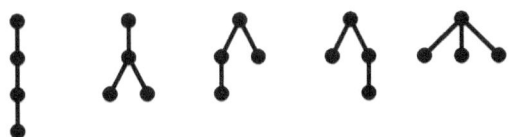

(ref: 4357)

Chapter 2: Bijection

Practice 12

Find the number of parallelograms in the following equilateral triangle of side length n which is made of some smaller unit equilateral triangles.

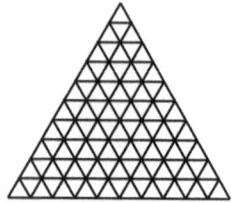

(ref: 4354)

Practice 13

Let $\mathbb{S} = \{1, 2, \cdots, 1000\}$ and $\mathbb{A}$ be a subset of $\mathbb{S}$. If the number of elements in $\mathbb{A}$ is 201 and their sum is a multiple of 5, then $\mathbb{A}$ is called *good*. How many good $\mathbb{A}$ are there?

(ref: 4367)

Practice 14

Assuming positive integer n satisfies $n \equiv 1 \pmod{4}$ and $n > 1$. Let $\mathbb{P} = \{a_1, a_2, \cdots, a_n\}$ be a permutation of $\{1, 2, \cdots, n\}$. If k_p denotes the largest index k associated with $\mathbb{P}$ such that the following inequality holds

$$a_1 + a_2 + \cdots + a_k < a_{k+1} + a_{k+2} + \cdots + a_n$$

Find the sum of k_p for all possible $\mathbb{P}$.

(ref: 4365)

Chapter 2: Bijection

Practice 15

Let S_n be the number of non-congruent triangles whose sides' lengths are all integers and circumferences equals n. Show that

$$S_{2n-1} - S_{2n} = \left\lfloor \frac{n}{6} \right\rfloor \quad \text{or} \quad \left\lfloor \frac{n}{6} \right\rfloor + 1$$

where $\lfloor x \rfloor$ returns the largest integer not exceeding the real number x.

(ref: 4424)

Practice 16

Let $x_i \in \{+1, -1\}$, $i = 1, 2, \cdots, 2n$. If their sum equals 0 and the following inequality holds for any positive integer k satisfying $1 \leq k < 2n$:

$$x_1 + x_2 + \cdots + x_k \geq 0$$

Find the number of possible ordered sequence $\{x_1, x_2, \cdots, x_{2n}\}$.

(ref: 4477)

Chapter 2: Bijection

Chapter 3

Recursion

3.1 Introduction

Some counting problems can be solved by establishing a recursion and then solving this relation. Using this method is appropriate when it is difficult to compute the result directly but a recursive relation is easier to be obtained.

The following steps are required to employ this method:

i) Establish a recursion.

ii) Determine initial values.

iii) Calculate the result manually or solve the recursion.

Let's review the last problem in 2010 AMC8 as an example.

Example 3.1.1

Everyday at school, Jo climbs a flight of 6 stairs. Joe can take the stairs 1, 2, or 3 at a time. For example, Jo could climb 3, then 1, then 2. In how many ways can Jo climb the stairs?

Solution

Let a_n be the number of different ways to climb n stairs. Because Jo can climb up to 3 steps at a time, there will be three possibilities for him to reach the nth step depending on the number of stairs in his last attempt:

i) If his last climb is one-step, then Jo must have stood on the $(n-1)^{th}$ stair before his last attempt. There are a_{n-1} ways for him to reach the $(n-1)^{th}$ step.

ii) If his last climb is two-step, then Jo must have stood on the $(n-2)^{th}$ stair before his last attempt. There are a_{n-2} ways for him to reach the $(n-2)^{th}$ step.

iii) If his last climb is three-step, then Jo must have stood on the $(n-3)^{th}$ stair before his last attempt. There are a_{n-3} ways for him to reach the $(n-3)^{th}$ step.

Therefore, the recursion is

$$a_n = a_{n-1} + a_{n-2} + a_{n-3}$$

Meanwhile, initial values can be determined as $a_1 = 1$, $a_2 = 2$, and $a_3 = 4$. It follows that

$$\begin{aligned} a_4 &= 4+2+1 &= 7 \\ a_5 &= 7+4+2 &= 13 \\ a_6 &= 13+7+4 &= \boxed{24} \end{aligned}$$

Done.

3.2 More Examples

The following example requires linear recursion to solve. The steps to solve such a sequence will not be discussed in detail here. They are covered in the book *Competition Algebra*.

Chapter 3: Recursion

Example 3.2.1

Find the total number of sequences of length n containing only letters A and B such that no two As are next to each other. For example, for $n = 2$, there are 3 possible sequences: AB, BA, and BB.

(ref: 4479)

Solution

Let's call a sequence without consecutive letters of A a good sequence, and F_n be the number of good sequences with n letters. Such a sequence can either start with letter A or letter B.

If it starts with A, then the 2nd letter must be B and the 3rd letter can be either A or B. In other words, if a good sequence of n letters starts with A, then its first two letters are fixed. Hence such a n-letter sequence is uniquely determined by its subsequence starting from the 3rd letter which has $(n-2)$ letters. This means that the number of such sequences with n-letter equals that of corresponding sequences having $(n-2)$ letters, i.e. F_{n-2}.

If it starts with B, then the 2nd letter can be either A or B. This means that the number of such sequences is the same as that of corresponding subsequence of $(n-1)$ letters, i.e. F_{n-1}.

It follows that
$$F_n = F_{n-1} + F_{n-2}$$
where $F_1 = 1$ and $F_2 = 2$. This is a standard linear regression and its solution is

$$F_n = \boxed{\frac{5+3\sqrt{5}}{10}\left(\frac{1+\sqrt{5}}{2}\right)^n + \frac{5-3\sqrt{5}}{10}\left(\frac{1-\sqrt{5}}{2}\right)^n}$$

Done.

Chapter 3: Recursion

Sometimes, it may be necessary to establish two related recursions in order to find the solution.

Example 3.2.2

Let S be a collection of $\{a_1, a_2, \cdots, a_n\}$ where every element $a_i \in \{1, 2, \cdots, k\}$. Find the number of S which has an even number of 1s.

(ref: 4416)

Solution

Let the number of S which has an even and an odd number of 1s be E_n and O_n respectively. Because there are totally k^n possible S, thus

$$E_n + O_n = k^n$$

Now consider the relation between an n-element S_n and an S_{n-1} of $(n-1)$ elements. If S_{n-1} has an odd number of 1s, then S_n needs to append a 1 in order to have an even number of 1. If S_{n-1} has an even number of 1, then S_n has $(k-1)$ choices to select an element not equal to 1. Therefore,

$$E_n = O_{n-1} + (k-1)E_{n-1}$$

Because $O_{n-1} + E_{n-1} = k^{n-1}$, this relation can be rewritten as

$$E_n = (k^{n-1} - E_{n-1}) + (k-1)E_{n-1} = (k-2)E_{n-1} + k^{n-1}$$

Meanwhile, it can be determined that $E_1 = k - 1$. So E_n equals

$$(k-2)E_{n-1} + k^{n-1}$$
$$= (k-2)((k-2)E_{n-2} + k^{n-2}) + k^{n-1}$$
$$= \cdots$$
$$= (k-2)^{n-1}(k-1) + (k-2)^{n-2}k + \cdots + (k-2)k^{n-2} + k^{n-1}$$
$$= \left((k-2)^{n-1} + (k-2)^{n-2}k + \cdots + (k-2)k^{n-2} + k^{n-1}\right) + (k-2)^n$$
$$= (k-2)^{n-1} \cdot \frac{1 - \left(\frac{k}{k-2}\right)^n}{1 - \frac{k}{k-2}} + (k-2)^n$$

Chapter 3: Recursion

$$= \boxed{\frac{k^n + (k-2)^n}{2}}$$

Done.

It is worth noting that recursion is not necessarily a closed form. It can be a bespoke system of relations. The following 2014 AMC12 problem offers such an example.

Example 3.2.3

In a small pond there are eleven lily pads in a row labeled 0 through 10. A frog is sitting on pad 1. When the frog is on pad N, $0 < N < 10$, it will jump to pad $(N-1)$ with probability $\frac{N}{10}$ and to pad $(N+1)$ with probability $\left(1 - \frac{N}{10}\right)$. Each jump is independent of the previous jumps. If the frog reaches pad 0 it will be eaten by a patiently waiting snake. If the frog reaches pad 10 it will exit the pond, never to return. What is the probability that the frog will escape without being eaten by the snake?

Solution

Let $P(N)$ be the frog's probability of survival when it is sitting on pad N. Then the desired answer is $P(1)$.

In the diagram below, numbered boxes denote the pads and the values above arrows show the probability when the frog jumps between corresponding pads.

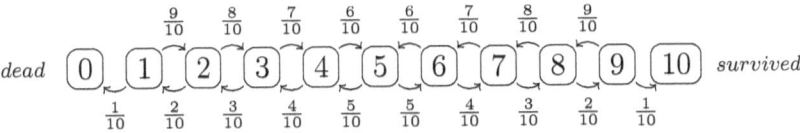

Chapter 3: Recursion

Observing the above diagram reveals that the probability transitions when $N \leq 5$ and those when $N \geq 5$ are symmetric. Hence, it must hold that $P(5) = \frac{1}{2}$.

When the frog sits on the Nth pad, it can jump to either the $(N-1)^{th}$ or the $(N+1)^{th}$ pad. The corresponding survival probabilities are $P(N-1)$ and $P(N+1)$, respectively. This means that the frog's current survival rate will be a weighted average of these two , i.e.:

$$P(N) = \frac{N}{10} \cdot P(N-1) + \left(1 - \frac{N}{10}\right) \cdot P(N+1)$$

It follows that

$$\begin{cases} P(0) = 0 \\ P(1) = \frac{1}{10}P(0) + \frac{9}{10}P(2) \\ P(2) = \frac{2}{10}P(1) + \frac{8}{10}P(3) \\ P(3) = \frac{3}{10}P(2) + \frac{7}{10}P(4) \\ P(4) = \frac{4}{10}P(3) + \frac{6}{10}P(5) \\ P(5) = \frac{1}{2} \end{cases}$$

Solving this system gives

$$P(1) = \boxed{\frac{63}{146}}$$

Done.

3.3 Practice

Practice 1

How many different ways are there to cover a 1×10 grid with some 1×1 and 1×2 pieces without overlapping?

(ref: 2495)

Practice 2

(Hanoi Tower) There are 3 identical rods labeled as A, B, C; and n disks of different sizes which can be slide onto any of these three rods. Initially, the n disks are stacked in ascending order of their sizes on A. What is the minimal number of moves in order to transfer all the disks to B providing that each move can only transfer one disk to another rod's topmost position and at no time, a bigger disk can be placed on top of a smaller one.

(ref: 4478)

Practice 3

Call a permutation $a_1, a_2, \ldots, a_n$ of the integers $1, 2, \ldots, n$ quasi-increasing if $a_k \leq a_{k+1} + 2$ for each $1 \leq k \leq n-1$. For example, 53421 and 14253 are quasi-increasing permutations of the integers 1, 2, 3, 4, 5, but 45123 is not. Find the number of quasi-increasing permutations of the integers $1, 2, \ldots, 7$.

(ref: 77 - AIME)

Chapter 3: Recursion

Practice 4

In the Banana Country, only Mr Decent always tells the truth and only Mr Joke always tells lies. Everyone else has a probability of p to tell a lie. One day, Mr Decent has decided to run for the President and told his decision to the first person who in turn told this to the second person. The second person then told this to the third person, and so on, till the nth person who told this news to Mr Joke. No one has been told this news twice in this process. Finally, Mr Joke announced Mr Decent's decision to everyone. What is the probability that Mr Joke's statement agrees with Mr Decent's intention?

(ref: 4418)

Practice 5

How many length ten strings consisting of only As and Bs contain neither "BAB" nor "BBB" as a substring?

(ref: 2046 - Exeter)

Practice 6

Dividing a circle into $n \geq 2$ sectors and coloring these sectors using $m \geq 2$ different colors. If no adjacent sectors can be colored the same, how many different color schemes are there?

(ref: 4407)

Practice 7

There are $2^{10} = 1024$ possible 10-letter strings in which each letter is either an A or a B. Find the number of such strings that do not have more than 3 adjacent letters that are identical.

(ref: 79 - AIME)

Practice 8

Let $S = \{a_1, a_2, \cdots, a_n\}$ be a permutation of $\{1, 2, \cdots, n\}$ which satisfies the condition that for every a_i, ($i = 1, 2, \cdots, n$), there exists an a_j where $i < j \leq n$ such that $a_j = a_i + 1$ or $a_j = a_i - 1$. Find the number of such S.

(ref: 4415)

Practice 9

Find a recursion for the last practice in the previous chapter (no need to solve the recursion for now):

Let $x_i \in \{+1, -1\}$, $i = 1, 2, \cdots, 2n$. If their sum equals 0 and the following inequality holds for any positive integer k satisfying $1 \leq k < 2n$:

$$x_1 + x_2 + \cdots + x_k \geq 0$$

Find the number of possible ordered sequence $\{x_1, x_2, \cdots, x_{2n}\}$.

Practice 10

Find the number of ways to divide a convex n-sided polygon into $(n-2)$ triangles using non-intersecting diagonals.

(ref: 4465)

Chapter 3: Recursion

Chapter 4

Integer Solution and Balls-in-Boxes

4.1 The Integer Solution Model

Some problems can be modeled as counting the number of integer solutions to the following equation

$$x_1 + x_2 + x_3 + \cdots + x_k = n \tag{4.1}$$

where k and n are two positive integers. Such problems have appeared more and more frequently in recent years. For instance:

Example 4.1.1

For some particular values of N, when $(a + b + c + d + 1)^N$ is expanded and like terms are combined, the resulting expression contains exactly 1001 terms that include all four variables $a, b, c,$ and d, each to some positive power. What is N?

(AMC10)

Chapter 4: Integer Solution and Balls-in-Boxes

This problem is related to the number of integer solutions to

$$x_1 + x_2 + x_3 + x_4 + x_5 = N$$

where x_1, x_2, x_3, $x_4 > 0$ and $x_5 \geq 0$. Explanation of the solutions to this examples and next few will be given in detail later.

Example 4.1.2

A parking lot has 16 spaces in a row. Twelve cars arrive, each of which requires one parking space, and their drivers chose spaces at random from among the available spaces. Auntie Em then arrives in her SUV, which requires 2 adjacent spaces. What is the probability that she is able to park?

(AMC12)

This problem is related to the number of integer solutions to

$$x_1 + x_2 + x_3 + x_4 + x_5 = 12$$

where x_1, $x_5 \geq 0$ and x_2, x_3, $x_4 > 0$.

Example 4.1.3

Let $\mathbb{A} = \{a_1, a_2, \cdots, a_{100}\}$ be a set containing 100 real numbers, $\mathbb{B} = \{b_1, b_2, \cdots, b_{50}\}$ be a set containing 50 real numbers, and f be a mapping from $\mathbb{A}$ to $\mathbb{B}$. Find the number of possible f if $f(a_1) \leq f(a_2) \leq \cdots f(a_1)$, and for every $b_i \in \mathbb{B}$, there exists an element $a_i \in \mathbb{A}$ such that the $f(a_i) = b_i$.

(China)

This problem is equivalent to finding the number of positive integer solutions to

$$x_1 + x_2 + \cdots + x_{50} = 100$$

The basic tools to count the integer solutions to *Equation 4.1* are the cut-the-rope method and the knives-and-balls model both

Chapter 4: Integer Solution and Balls-in-Boxes

of which are discussed in the first *Counting* book. Some textbooks call these two methods the stars-and-bars method. These names are largely interchangeable because they share the same approach. These techniques can be used to count the solutions directly when all the variables are positive or non-negative. These two cases are the simplest and also stepping stones for solving more complex problems.

Meanwhile, this integer solution model can be generalized in several different ways.

One way focuses on the restrictions of these variables x_i. For instance, both *Example 4.1.1* and *4.1.2* mentioned earlier involve a mix of positive and non-negative integers. Such scenarios can still be tackled by tweaking the two basic techniques. However, there are other variations which will need more powerful methods. Some of such examples will be given in *Section 4.3* later. A better technique to solve these problems is generating function which will also be discussed in this book.

Additionally, this integer solution pattern can be regarded as a special case of the balls-in-boxes model. The latter is to count the number of ways to put n balls into k boxes. Depending on factors such as whether balls are distinguishable or not, boxes are distinguishable or not, empty boxes are permitted or not, etc, the balls-in-boxes model can be transformed into several different sub-models. One of them is the integer solution model. Different sub models' complexities differ significantly. Some of them will be covered later in this chapter.

4.2 Count Integer Solutions

Two basic cases are when all the variables are (1) positive and (2) non-negative. Given the equation

$$x_1 + x_2 + x_3 + \cdots + x_k = n \tag{4.2}$$

Chapter 4: Integer Solution and Balls-in-Boxes

where both n and k are positive integers, there are

$$\binom{n-1}{k-1} \text{ positive integer solutions}$$

and

$$\binom{n+k-1}{k-1} \text{ non-negative integer solutions}$$

The first conclusion can be explained using the cut-the-rope method with n balls as shown below.

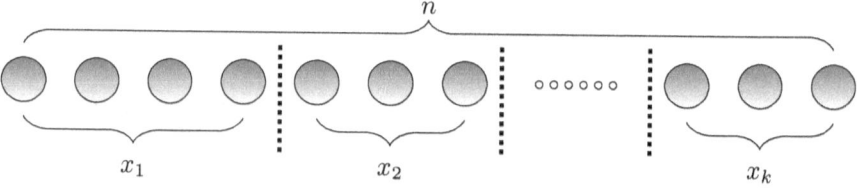

One positive integer solution to (4.2) is equivalent to one way of dividing n balls into k non-empty sets. The number of balls in each set is the value of each variable x_i. Dividing n balls into k non-empty sets is the same as placing $(k-1)$ cuts chosen from $(n-1)$ intervals between balls. Hence, the result is $\binom{n-1}{k-1}$.

Similarly, the knives-and-balls method can be used count non-negative integer solutions. One non-negative integer solution to (4.1) is equivalent to finding $(k-1)$ places to put $(k-1)$ knives chosen from $(n+k-1)$ spots. Then, the n balls can be put at the remaining n places. The number of balls between adjacent knives is the value of corresponding x_i. If two knives are next to each other without any ball in between, then the corresponding variable will equal 0.

It is worth noting that these two explanation can be both viewed

Chapter 4: Integer Solution and Balls-in-Boxes

as applications of the bijection method where problems are modeled with equivalent patterns whose solutions are known.

Example 4.2.1

Find the number of positive integer solutions to the equation

$$x_1 + x_2 + x_3 = 5$$

Solution

The answer is $\binom{5-1}{3-1} = \boxed{6}$. This result can be verified by manually listing all the solutions:

(1, 1, 3), (1, 2, 2), (1, 3, 1), (2, 1, 2), (2, 2, 1), (3, 1, 1)

Done.

Example 4.2.2

How many non-negative integer triplets (x_1, x_2, x_3) satisfies the relation

$$x_1 + x_2 + x_3 = 5$$

Solution

The answer is $\binom{5+3-1}{3-1} = \boxed{21}$. All the solutions are

(0, 0, 5), (0, 1, 4), (0, 2, 3), (0, 3, 2), (0, 4, 1), (0, 5, 0),
(1, 0, 4), (1, 1, 3), (1, 2, 2), (1, 3, 1), (1, 4, 0),
(2, 0, 3), (2, 1, 2), (2, 2, 1), (2, 3, 0),
(3, 0, 2), (3, 1, 1), (3, 2, 0),
(4, 0, 1), (4, 1, 0),
(5, 0, 0)

Done.

Chapter 4: Integer Solution and Balls-in-Boxes

Example 4.2.3

Find the number of integer solutions to the equation

$$x_1 + x_2 + x_3 = 5$$

where $x_1 \geq 0$ and $x_2, x_3 > 0$.

Such a problem where some variables are positive and some are non-negative is a common variation to the two basic cases discussed so far. This type of problems can be solved by transforming the given equations to either of the two basic cases using variable substitution.

Solution

Let $x_1' = x + 1$. Then x_1' is a positive integer. It follows that the original problem is equivalent to counting the positive integer solutions to

$$x_1' + x_2 + x_3 = 6$$

Hence, the answer is $\binom{6-1}{3-1} = \boxed{10}$. This result can be verified by listing all the solutions:

(0, 1, 4), (0, 2, 3), (0, 3, 2), (0, 4, 1)
(1, 1, 3), (1, 2, 2), (1, 3, 1),
(2, 1, 2), (2, 2, 1),
(3, 1, 1),

It is also possible to transform this problem to the case of where all variables are non-negative integers by replacing x_2 and x_3.

Done.

Chapter 4: Integer Solution and Balls-in-Boxes

Example 4.2.4

A parking lot has 16 spaces in a row. Twelve cars arrive, each of which requires one parking space, and their drivers chose spaces at random from among the available spaces. Auntie Em then arrives in her SUV, which requires 2 adjacent spaces. What is the probability that she is able to park?

(ref: 790 - AMC12)

Solution

Instead of investigating when Auntie Em can park her car, let's examine when she cannot. Such situations arise when all available parking spaces are separated by one or more vehicles.

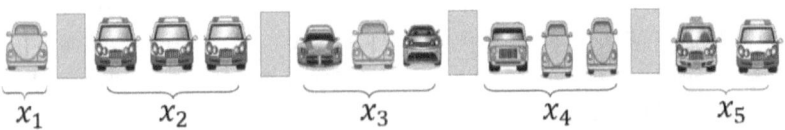

The diagram above shows that four spaces are separated by one or more cars. Each variable, x_i, denotes the number of cars in that block. There are totally 12 cars. Hence these variables should add up to 12:

$$x_1 + x_2 + x_3 + x_4 + x_5 = 12 \qquad (4.3)$$

Among these variables, x_2, x_3, and x_4 must be positive because otherwise two vacant spaces will join each other. But x_1 and x_5 can be zero in which case a vacant space happens to be located at the start or end of the parking lot.

Hence, the number of cases when Auntie Em cannot park her car is the number of integer solutions to (4.3) where $x_1, x_5 \geq 0$ and $x_2, x_3, x_4 > 0$. Using the techniques presented in this section gives the answer as $\binom{13}{4}$.

Chapter 4: Integer Solution and Balls-in-Boxes

It follows that the final answer to the given problem is

$$1 - \frac{\binom{13}{4}}{\binom{16}{12}} = \boxed{\frac{17}{28}}$$

where $\binom{16}{12}$ is the total number of ways to park 12 vehicles in a 16-spot lot.

Done.

4.3 Other Situations

Seemingly minor differences may be significant. There are situations where problems appear to be similar to the ones discussed in this chapter but are more difficult to solve. One such example is given in the first *Counting* book.

Example 4.3.1

Roll a standard 6-sided die three times. What is the probability that their sum is 8?

Because the total number of outcomes is 6^3, the key so solve this problem is to count the outcomes when the sum of three rolls equal 8. This count is the same as the number of integer solutions to

$$x_1 + x_2 + x_3 = 8 \tag{4.4}$$

subject to the condition that $1 \leq x_1, x_2, x_3 \leq 6$. Though this problem appears similar to the examples given in the proceeding section, it is more difficult to be tackled. The solution given in the first book exploits the fact that counting unqualified outcomes manually is easy in this particular situation. Hence, that approach is not a general solution to this type of problems. One general way to solve

Chapter 4: Integer Solution and Balls-in-Boxes

such problems is to use the generating function method which will be discussed later in the book.

Example 4.3.2

Find the number of positive integer solutions to the equation

$$x_1 + 2x_2 + 5x_3 = 20$$

This is another type of problems that seems similar but actually is different. If the involved number is small like in this example, a workable method is just to manually list all the qualified solutions. A general solving method to such problems is generating function.

Example 4.3.3

How many different ways are there to pay $20 using $1, $2, and $5 notes?

This problem is the same as the previous example except that x_1, x_2, and x_3 are non-negative integers.

Example 4.3.4

How many different ways are there to express 20 as the sum of 1, 2, and 5?

Expressing an integer as the sum of several smaller positive integers is called number partition. Depending on whether order matters or not, i.e. whether $11 + 9$ is treated as the same as $9 + 11$, a number partition problem may be modeled using *Example 4.3.2* or the balls-in-boxes pattern. There is no known closed form solution to a general number partition problem, but some special cases can be handled by manual counting or other techniques such as the generating function method.

Chapter 4: Integer Solution and Balls-in-Boxes

4.4 The Balls-in-Boxes Model

This model counts the number of ways to put n balls into k boxes. There are three main parameters influencing the solutions:

i) whether empty boxes are permitted or not

ii) whether balls are distinguishable or not

iii) whether boxes are distinguishable or not

Different combinations of parameters result in different solutions. There are a total of 8 variations. Some of them have already been covered in the first *Counting* book and elsewhere in this book. This section aims to provide a brief summary.

The simplest case is when both balls and boxes are distinguishable, and empty boxes are permitted. In this case, each of these n balls have k choices of boxes to be placed in. Therefore, the number of ways is k^n.

When empty boxes are not permitted, then those cases when at least one box is empty must be excluded from the previous result. The correct result can be computed using the inclusion-and-exclusion principle:

$$k^n - \binom{k}{1}(k-1)^n + \binom{k}{2}(k-2)^n - \binom{n}{3}(k-3)^n + \cdots \qquad (4.5)$$

When balls are indistinguishable but the boxes are distinguishable, then this model is equivalent to counting the integer solutions to the equation

$$x_1 + x_2 + \cdots + x_k = n$$

where the value of x_i indicates the number of balls in the ith box. If no empty box is allowed, then all variables must be positive integers. Otherwise, if empty boxes are permitted, then all the variables are non-negative integers.

Chapter 4: Integer Solution and Balls-in-Boxes

The cases when boxes are indistinguishable will be generally much more challenging.

For example, in order to understand the model when balls are distinguishable but boxes are not, let's consider an example from the first book: how many ways to equally distribute 6 different books (distinguishable balls) to 3 piles (indistinguishable boxes)? The answer to this particular problem is

$$\frac{1}{3!}\binom{6}{2}\binom{4}{2}\binom{2}{2}$$

The divisor 3! above is the duplicate count. Dividing it is necessary because the boxes are indistinguishable. However, the duplicate count may differ when the numbers of books in these piles changes. Consequently, it will be complicated to compute the total possible distributions, especially when empty box is permitted.

If both balls and boxes are indistinguishable, then this model is equivalent to expressing n as the sum of up to k positive integers when order does not matter. When empty box is not permitted, exactly k numbers are required. Otherwise less than k numbers are permitted. Such type of problems will unlikely appear in high school competitions except those which are specifically designed and can be solved in special ways.

Chapter 4: Integer Solution and Balls-in-Boxes

4.5 Practice

Practice 1

Find the number of positive integer solutions to the following equation:
$$x_1 + x_2 + \cdots + x_5 = 14$$

(ref: 2526)

Practice 2

Find the number of non-negative integer solutions to the following equation:
$$x_1 + x_2 + \cdots + x_5 = 14$$

(ref: 4134)

Practice 3

Find the number of integer solutions to the following equation:
$$x_1 + x_2 + \cdots + x_6 = 12$$

where $x_1, x_5 \geq 0$ and $x_2, x_3, x_4 > 0$

(ref: 4135)

Practice 4

Find the number of non-negative integer solutions to the equation
$$2x_1 + x_2 + x_3 + \cdots + x_9 + x_{10} = 3$$

(ref: 4409)

Chapter 4: Integer Solution and Balls-in-Boxes

Practice 5

Pat is to select six cookies from a tray containing only chocolate chip, oatmeal, and peanut butter cookies. There are at least six of each of these three kinds of cookies on the tray. How many different assortments of six cookies can be selected?

(ref: 2824 - AMC10)

Practice 6

Find the number of ordered quadruples of integer (a, b, c, d) satisfying $1 \leq a < b < c < d \leq 10$.

(ref: 4276)

Practice 7

How many different outcomes are there if three dices are rolled at the same time?

(ref: 4419)

Practice 8

A total of 2018 tickets, numbered 1, 2, 3, $\cdots$, 2017, 2018 are placed in an empty bag. Alfred removes ticket a from the bag. Bernice then removes ticket b from the bag. Finally, Charlie removes ticket c from the bag. They notice that $a < b < c$ and $a + b + c = 2018$. In how many ways could this happen?

(ref: 2679)

Chapter 4: Integer Solution and Balls-in-Boxes

Practice 9

Find the number of non-decrease sequences of length n and each element is a non-negative integer not exceeding d.

(ref: 4145)

Practice 10

Let $\mathbb{A} = \{a_1, a_2, \cdots, a_{100}\}$ be a set containing 100 real numbers, $\mathbb{B} = \{b_1, b_2, \cdots, b_{50}\}$ be a set containing 50 real numbers, and $\mathcal{F}$ be a mapping from $\mathbb{A}$ to $\mathbb{B}$. Find the number of possible $\mathcal{F}$ if $\mathcal{F}(a_1) \leq \mathcal{F}(a_2) \leq \cdots \mathcal{F}(a_1)$, and for every $b_i \in \mathbb{B}$, there exists an element $a_i \in \mathbb{A}$ such that the $\mathcal{F}(a_i) = b_i$.

(ref: 4361 - China)

Practice 11

How many ordered integers (x_1, x_2, x_3, x_4) are there such that $0 < x_1 \leq x_2 \leq x_3 \leq x_4 < 7$?

(ref: 4411)

Practice 12

Define an ordered quadruple of integers (a, b, c, d) as interesting if $1 \leq a < b < c < d \leq 10$, and $a + d > b + c$. How many interesting ordered quadruples are there?

(ref: 260 - AIME)

Chapter 4: Integer Solution and Balls-in-Boxes

Practice 13

How many ways are there to arrange 8 girls and 25 boys to sit around a table so that there are at least 2 boys between any pair of girls? If a sitting plan can be simply rotated to match another one, these two are treated as the same.

(ref: 4366 - China)

Practice 14

Six men and some number of women stand in a line in random order. Let p be the probability that a group of at least four men stand together in the line, given that every man stands next to at least one other man. Find the least number of women in the line such that p does not exceed 1 percent.

(ref: 251 - AIME)

Practice 15

How many different ways to write a positive integer n as a sum of m different positive integers? Different sequences are treated as distinct.

(ref: 4413)

Chapter 4: Integer Solution and Balls-in-Boxes

Chapter 5

Combinatorial Identities

5.1 Binomial / Multinomial Theorem

The binomial theorem is often referred as binomial expansion:

Theorem 5.1.1 Binomial Theorem

$$(a+b)^n = \sum_{k=0}^{n} \binom{n}{k} a^{n-k} b^k \qquad (5.1)$$

When there are more than two variables involved, the binomial theorem becomes multinomial theorem:

Theorem 5.1.2 Multinomial Theorem

$$(x_1 + x_2 + \cdots + x_k)^n = \sum_{p_1+p_2+\cdots+p_k=n} \binom{n}{p_1, p_2, \cdots, p_k} x_1^{p_1} x_2^{p_2} \cdots x_k^{p_k} \qquad (5.2)$$

where $\sum_{p_1+p_2+\cdots+p_k=n}$ means to sum all the cases where the condition $p_1 + p_2 + \cdots + p_k = n$ is met, and

$$\binom{n}{p_1, p_2, \cdots, p_k} = \frac{n!}{p_1! \cdot p_2! \cdots p_k!}$$

where all of $p_1, p_2, \cdots, p_k$ are non-negative integers.

It can be verified that (5.1) is a special case of (5.2) because where there are just two variables, multinomial expansion becomes

$$(x_1 + x_2)^n = \sum_{p_1+p_2=n} \binom{n}{p_1, p_2} x_1^{p_1} x_2^{p_2}$$

Replacing p_1 with k and noting $p_2 = n - p_1 = n - k$ in the above relation leads to (5.1) immediately.

Meanwhile, applying the non-negative integer solutions pattern gives the number of situations satisfying $p_1 + p_2 + \cdots + p_k = n$ as

$$\binom{n+k+1}{k-1}$$

This is the number of terms when (5.2) is expanded and like terms are merged. When $k = 2$, the above equation equals $(n+1)$. This agrees with the fact that are are $(n+1)$ terms in the expanded form of $(a+b)^n$.

5.2 The Counting Method

Both the binomial and multinomial theorems can be explained using the counting method. The counting method is to construct a counting problem which can be solved in two different ways. Each way produces a result in one expression. Because both of these two expressions are answers to the same problem, they must equal.

In order to prove the binomial theorem, it is sufficient to show that the coefficient of the $a^{n-k}b^k$ term after expanding the left side of *(5.1)* equals to the coefficient of the same term on the right side. This is because two polynomials will be equal if all the corresponding coefficients are equal.

The left side of *(5.1)* can be rewritten as

$$\underbrace{(a+b)(a+b)\cdots(a+b)}_{n}$$

Every term in the form of $a^{n-k}b^k$ must be the result of choosing a from $(n-k)$ of these n brackets, and b from the remaining k brackets. After all these $a^{n-k}b^k$ terms are merged, the coefficient of the merged term must equal the number of ways of selecting $(n-k)$ brackets for a and k brackets for b from a total of n choices. Using the basic counting technique gives the answer as $\binom{n}{n-k} = \binom{n}{k}$.

This line of reasoning can also be described in a more counting like manner. Consider this counting problem: given n boxes each of which contains one white and one black balls, how many different ways are there to retrieve one ball from each box such that a total of $(n-k)$ white balls and k black balls are obtained?

A similar counting problem can be used to explain the multinomial theorem: There are n boxes each of which contains exactly same set of balls. Each set has k balls in k different colors. One ball is retrieved from each of these n boxes. How many different ways are there to retrieve p_1 balls of the first color, p_2 balls of the second color, ..., and p_k balls of the kth color? Using the basic counting techniques discussed in the first book, the answer is

$$\frac{n!}{p_1! \cdot p_2! \cdots p_k!}$$

which is exactly the definition of $\binom{n}{p_1, p_2, \cdots, p_k}$.

Let's consider another example which can be proved using the counting method.

Chapter 5: Combinatorial Identities

> **Theorem 5.2.1   Vandermonde's Identity**
>
> $$\binom{m+n}{r} = \binom{m}{0}\binom{n}{r} + \binom{m}{1}\binom{n}{r-1} \cdots + \binom{m}{r}\binom{n}{0} \qquad (5.3)$$

Proof

Let's count the number of shortest paths along the grid from point A to B in a $(m+n-r) \times r$ rectangular grid.

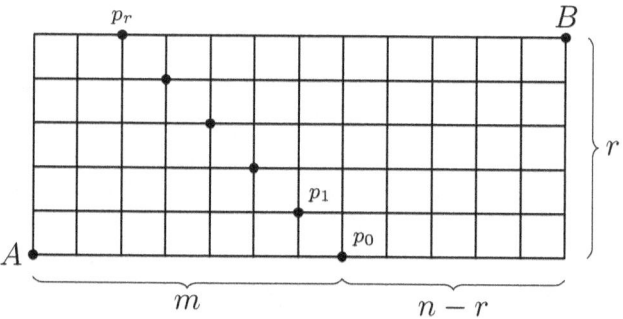

On one hand, applying the basic pattern gives the answer as

$$\binom{m+(n-r)+r}{r} = \binom{m+n}{r}$$

On the other hand, let's consider some mid-way points, p_0, p_1, $\cdots$, p_r as shown. (These points are on the line $x+y=m$.) Any path from A to B must pass one and only one of these points.

There are $\binom{m}{0}$ ways from A to p_0 and $\binom{(n-r)+r}{r} = \binom{n}{r}$ ways from p_0 to B. Therefore, there is a total of $\binom{m}{0}\binom{n}{r}$ ways from A to B via p_0. Similarly, there are $\binom{m}{1}\binom{n}{r-1}$ ways via p_1, and so on. Hence, the total number of paths from A to B must be

$$\binom{m}{0}\binom{n}{r} + \binom{m}{1}\binom{n}{r-1} \cdots + \binom{m}{r}\binom{n}{0}$$

Chapter 5: Combinatorial Identities

They are two different approaches to calculate the same quantity, thus must be equal, i.e.:

$$\binom{m+n}{r} = \binom{m}{0}\binom{n}{r} + \binom{m}{1}\binom{n}{r-1} \cdots + \binom{m}{r}\binom{n}{0}$$

QED

5.3 The Coefficient Method

Another way to prove the Vandermonde's identity is to employ the coefficient method. This method uses two different approaches to calculate the coefficient of the same term. Because the two results must equal, therefore the corresponding expressions must be equal too. Its idea is already used in explaining the binomial theorem. In that case, a polynomial identity, i.e. the binomial expansion itself (5.1), is already given. However, in a typical case when the coefficient method is employed, the key is to find a suitable polynomial identity which is usually not given.

Example 5.3.1

Use the coefficient method to prove

$$\binom{m+n}{r} = \binom{m}{0}\binom{n}{r} + \binom{m}{1}\binom{n}{r-1} \cdots + \binom{m}{r}\binom{n}{0}$$

Proof

Consider the coefficient of the term x^r in the following polynomial identity.

$$(1+x)^{m+n} = (1+x)^m (1+x)^n$$

On the left side, it is $\binom{m+n}{r}$ by the binomial theorem.

Chapter 5: Combinatorial Identities

Meanwhile, the right side can be expanded to

$$\left(\binom{m}{0} + x\binom{m}{1} + \cdots x^m\binom{m}{m}\right)\left(\binom{n}{0} + x\binom{n}{1} + \cdots x^n\binom{n}{n}\right)$$

An x^r term in the further expanded form must be a product of x^k from the first bracket and x^{r-k} from the second bracket, where $0 \le k \le r$. Accordingly, its coefficient will be the $\binom{m}{k}\binom{n}{r-k}$. It follows that the final coefficient of x^r after merging must equal

$$\sum_{k=0}^{r}\binom{m}{k}\binom{n}{r-k} = \binom{m}{0}\binom{n}{r} + \binom{m}{1}\binom{n}{r-1} \cdots + \binom{m}{r}\binom{n}{0}$$

Because these two results refer to the same quantity, so they must equal, i.e.

$$\binom{m+n}{r} = \binom{m}{0}\binom{n}{r} + \binom{m}{1}\binom{n}{r-1} \cdots + \binom{m}{r}\binom{n}{0}$$

QED

The coefficient method is often used in proving combinatorial identity and also is widely used in the generating function method.

5.4 Pascal Triangle

Pascal triangle, as shown below, is an infinite number of positive integers arranged in a triangular form. The two outermost edges are all 1s and a number in the inner area is always the sum of the two integers immediately above it.

Chapter 5: Combinatorial Identities

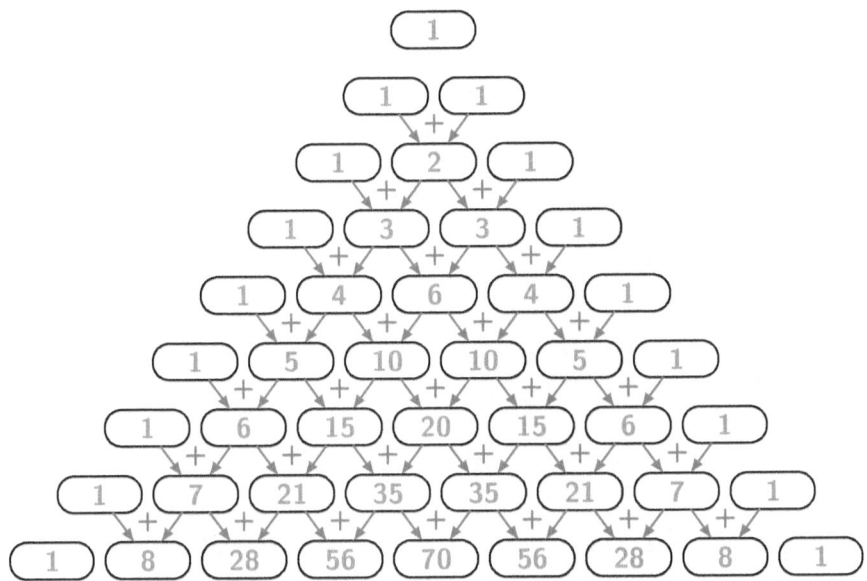

By convention, the top row is designated as the 0th row and the first entry in each row is said to be the 0th element.

Pascal triangle has a close relationship with combination numbers. For example, the kth element in the nth row equals $\binom{n}{k}$. Then, the way Pascal triangle is constructed can be expressed using the following combinatorial identity for any positive integer n and k. This relation is often referred as the Pascal triangle identity.

$$\binom{n}{k-1} + \binom{n}{k} = \binom{n+1}{k} \qquad (5.4)$$

Another related combinatorial relation is that the sum of all elements in the nth row equals 2^n. For example, the sum of the 0th row at the top is $1 = 2^0$. The sum of the 1th row equals $1 + 1 = 2^1$. The next row is $1+2+1 = 2^2$. This property can be written as

$$\binom{n}{0} + \binom{n}{1} + \cdots + \binom{n}{n} = 2^n$$

Chapter 5: Combinatorial Identities

Another interesting property is stated below.

> Let k be the number of 1s in the binary representation of a positive integer n. Then the nth row in Pascal triangle has 2^k odd numbers.

For example, because $0_{10} = 0_2$ has no one, there is $2^0 = 1$ odd number on the top row. Similarly, there are $2^1 = 2$ odd numbers in the $1_{10}^{st} = 1_2$ row, $2^1 = 1$ odd numbers in the $2_{10}^{nd} = 10_2$ row, $2^2 = 4$ odd numbers in the $3_{10}^{rd} = 11_2$ row, and so on.

This property can be generalized as

> Let n be a positive integer and its binary representation be $b_t b_{t-1} \cdots b_1 b_0$. Then it holds that
>
> $$\binom{n}{2^j} \equiv b_j \pmod{2} \tag{5.5}$$
>
> where j is a non-negative integer and we define $\binom{n}{m} = 0$ if $m > n$.

5.5 Modular Properties

When a combinatorial expression appears in a number theory modular related problem, the following property is often involved.

Example 5.5.1

Let p be a prime, and k be a positive integer satisfying $0 < k < p$. Show that $\binom{p}{k}$ is a multiple of p.

Chapter 5: Combinatorial Identities

Proof

Because p is a prime, therefore it is co-prime to any positive integer smaller than it. Then, in the definition

$$\binom{p}{k} = \frac{p(p-1)\cdots(p-k+1)}{1\cdot 2 \cdots k}$$

all of the factors in the denominator are co-prime to p. This means the p in the numerator will not be cancelled or reduced by any of the factors in the denominator. Meanwhile, $\binom{p}{k}$ is an integer. Hence, it must be a multiple of p.

QED

The conclusion of this example can lead to the following relation whose proof will be left as a practice problem.

> For any positive integer k and prime p, it must hold that
>
> $$(1+k)^p \equiv 1 + k^p \pmod{p} \qquad (5.6)$$

5.6 The Special Value Method

The special value method assigns specific numerical values to the binomial expansion formula as a way to derive various desired results. For this purpose, *(5.1)* on *page 41* is often written as the following for simplicity.

$$(1+x)^n = \sum_{k=0}^{n} \binom{n}{k} x^k = \binom{n}{0} + \binom{n}{1} x + \cdots + \binom{n}{n} x^n \qquad (5.7)$$

The following two well-known results can both be derived using the special value method.

Chapter 5: Combinatorial Identities

$$\binom{n}{0} + \binom{n}{1} + \binom{n}{2} + \cdots + \binom{n}{n} = 2^n \qquad (5.8)$$

This identity can be derived by setting $x = 1$ in (5.7).

$$\binom{n}{0} + \binom{n}{2} + \binom{n}{4} + \cdots = \binom{n}{1} + \binom{n}{3} + \binom{n}{5} + \cdots = 2^{n-1} \qquad (5.9)$$

Setting $x = -1$ to (5.7) gives

$$\binom{n}{0} - \binom{n}{1} + \binom{n}{2} - \binom{n}{3} + \binom{n}{4} - \binom{n}{5} + \cdots = 0 \qquad (5.10)$$

Then (5.8) + (5.10) can lead to

$$\binom{n}{0} + \binom{n}{2} + \binom{n}{4} + \cdots = 2^{n-1}$$

And (5.8) − (5.10) gives

$$\binom{n}{1} + \binom{n}{3} + \binom{n}{5} + \cdots = 2^{n-1}$$

Not only integers can be used as special values, but also fractions, even complex numbers can be used too.

Example 5.6.1

Let $a_n = \binom{2020}{3n-1}$. Find the vale of $\sum_{n=1}^{673} a_n$.

This problem is very similar to (5.8) and (5.9) except the interval now is 3 in the target combinatorial sequence. When the interval is 1 as in the case of (5.8), a single special value of 1 is used which is the root to $x^1 = 1$. When the interval becomes 2 in (5.9), two special values ±1 are used which are two roots to the equation $x^2 = 1$. Thus, it is natural to try three roots to $x^3 = 1$ in solving this problem.

Chapter 5: Combinatorial Identities

Solution

Let 1, ω, and ω^2 be the three distinct roots to the equation $x^3 = 1$ where $\omega = e^{i\frac{2\pi}{3}}$. Then we have $\omega^3 = 1$. Setting these three roots to *(5.7)*, respectively leads to

$$2^n = \binom{n}{0} + \binom{n}{1} + \binom{n}{2} + \binom{n}{3} + \cdots \qquad (5.11)$$

$$(1+\omega)^n = \binom{n}{0} + \binom{n}{1}\omega + \binom{n}{2}\omega^2 + \binom{n}{3} + \cdots \qquad (5.12)$$

$$(1+\omega^2)^n = \binom{n}{0} + \binom{n}{1}\omega^2 + \binom{n}{2}\omega + \binom{n}{3} + \cdots \qquad (5.13)$$

In order to extract the sequence of $\binom{n}{2}$, $\binom{n}{5}$, $\cdots$, coefficients of these terms need to be 1. Also note that

$$1 + \omega + \omega^2 = 0 \implies 1 + \omega = -\omega^2, \; 1 + \omega^2 = -\omega$$

Hence, *(5.11)* + *(5.12)* $\cdot \omega$ + *(5.13)* $\cdot \omega^2$ gives

$$2^n + (-1)^n \omega^{2n+1} + (-1)^n \omega^{n+2} = 3 \times \left(\binom{n}{2} + \binom{n}{5} + \binom{n}{8} + \cdots \right)$$

Therefore

$$\sum_{k=1}^{\lfloor \frac{n}{3} \rfloor} \binom{n}{3k-1} = \frac{1}{3} \cdot \left(2^n + (-1)^n \omega^{2n+1} + (-1)^n \omega^{n+2} \right)$$

Setting $n = 2020$ to the previous relation yields the final result:

$$\sum_{k=1}^{\infty} \binom{n}{3k-1} = \frac{1}{3} \cdot \left(2^{2020} + (-1)^{2020} \omega^{4041} + (-1)^{2020} \omega^{2022} \right)$$

$$= \frac{1}{3} \cdot \left(2^{2020} + 1 + 1 \right)$$

$$= \boxed{\frac{1}{3} \cdot \left(2^{2020} + 2 \right)}$$

Done.

If the sum of another sequence is required, such as $\binom{n}{0} + \binom{n}{3} +$

Chapter 5: Combinatorial Identities

$\binom{n}{6} + \cdots$ or $\binom{n}{1} + \binom{n}{4} + \binom{n}{7} + \cdots$, then *(5.11)*, *(5.12)*, and *(5.13)* need to be multiplied by different coefficients before being summed up. It is worth pointing out that though complex number is used in the process, the final result is always an integer.

It turns out that this method can be applied to a more general case. Let $\{a_i\}$, $(i = 0, 1, 2, \cdots)$ be any sequence, not limited to the binomial expansion coefficients, and

$$f(x) = a_0 + a_1 x + a_2 x^2 + a_3 x^3 + \cdots$$

then the partial sum

$$S_{k,d} = a_k + a_{k+d} + a_{k+2d} + a_{k+3d} + \cdots$$

where the interval d is a positive integer and the initial index k is a non-negative integer less than d, can be computed using the special value method as descried in *Example 5.6.1*. The result is

$$S_{k,d} = \frac{1}{d} \sum_{j=0}^{d-1} \left(\omega^{k(d-j)} f(\omega^j) \right) \qquad (5.14)$$

where $\omega = e^{i \frac{2\pi}{d}}$ is a complex root to equation $x^d = 1$.

Example 5.6.1 can also be solved directly using *(5.14)* by setting $d = 3$ and $k = 2$:

$$\begin{aligned}
S_{2,3} &= \frac{1}{3} \sum_{j=0}^{2} \left(\omega^{2 \cdot (3-j)} f(\omega^k) \right) \\
&= \frac{1}{3} \left(f(1) + \omega f(\omega) + \omega^2 f(\omega^2) \right) \\
&= \frac{1}{3} \left(2^{2020} + \omega (1+\omega)^{2020} + \omega^2 (1+\omega^2)^{2020} \right) \\
&= \frac{1}{3} \left(2^{2020} + \omega (-\omega^2)^{2020} + \omega^2 (-\omega)^{2020} \right) \\
&= \frac{1}{3} \left(2^{2020} + 1 + 1 \right)
\end{aligned}$$

$$= \frac{1}{3}\left(2^{2020} + 2\right)$$

Additionally, when d is odd, then it holds that

$$1 + x^d = (1+x)(1+\omega x)(1+\omega^2 x)\cdots(1+\omega^{d-1}x) \quad (5.15)$$

This relation holds because both sides are d-degree polynomials having the same set of d distinct roots: $-1, -\omega, -\omega^2, \cdots, -\omega^{d-1}$ and the same constant term of 1.

5.7 Hockey Stick Identity

Hockey stick identity appears frequently in middle level competitions. It states that

$$\binom{k}{k} + \binom{k+1}{k} + \binom{k+2}{k} + \cdots + \binom{n}{k} = \binom{n+1}{k+1} \quad (5.16)$$

This identity can be visualized on Pascal triangle which is illustrated below. The sum of a series of entries starting on one outermost edge and running down parallel to the other outermost edge equals to the entry just below the last entry which is not on the running path. The shape of such a running series with a turn at the end resembles the shape of a hockey stick. This is where the name of this identity comes from.

Chapter 5: Combinatorial Identities

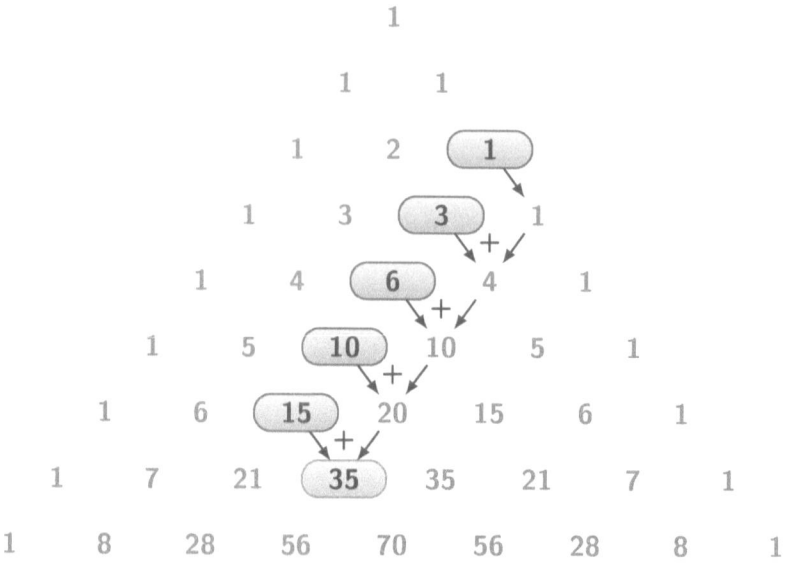

The illustration above shows

$$1 + 3 + 6 + 10 + 15 = 35$$

This equation can be written in the combinatorial form as

$$\binom{2}{2} + \binom{3}{2} + \binom{4}{2} + \binom{5}{2} + \binom{6}{2} = \binom{7}{3}$$

The proof of hockey stick identity can also be visualized on Pascal triangle. Because all entries on an outermost edge are 1, thus shifting the starting point one position downward will not change the sum. Then it is possible to repeatedly replace the two shoulder entries with their sum node below until reaching the end of the running series. This is also shown in the diagram above. Accordingly, the proof can be expressed in a more formal way using the Pascal triangle identity *(5.4)* on *page 47* as

$$\binom{k}{k} + \binom{k+1}{k} + \binom{k+2}{k} + \cdots + \binom{n}{k}$$
$$= \binom{k+1}{k+1} + \binom{k+1}{k} + \binom{k+2}{k} + \cdots + \binom{n}{k}$$
$$= \binom{k+2}{k+1} + \binom{k+2}{k} + \cdots + \binom{n}{k}$$

Chapter 5: Combinatorial Identities

$$= \cdots$$
$$= \binom{n}{k+1} + \binom{n}{k}$$
$$= \binom{n+1}{k+1}$$

Hockey stick identity is often used to simplify a combinatorial expression obtained during problem solving. Here is an example from AIME.

Example 5.7.1

Consider all 1000-element subsets of the set $\{1, 2, 3, ..., 2015\}$. From each such subset choose the least element. Find the arithmetic mean of all of these least elements.

(AIME)

Solution

There are totally $\binom{2015}{1000}$ subsets. For a subset to have k be the least number, the remaining 999 numbers must be selected from $(k+1)$ to 2015. There are $\binom{2015-k}{999}$ such subsets. Thus, the desired result is

$$M = \underbrace{\left(1 \cdot \binom{2014}{999} + 2\binom{2013}{999} + \cdots + 1016\binom{999}{999}\right)}_{A} \div \binom{2015}{1000}$$

The value of A can be computed by applying hockey stick identity multiple times. This is shown below:

$$A = 1 \cdot \binom{2014}{999} + 2 \cdot \binom{2013}{999} + \cdots + 1016 \cdot \binom{999}{999}$$
$$= \left(\binom{2014}{999} + \binom{2013}{999} + \cdots + \binom{999}{999}\right)$$
$$+ \left(\binom{2013}{999} + \cdots + \binom{999}{999}\right)$$
$$+ \cdots$$
$$+ \binom{9999}{9999}$$

55

Chapter 5: Combinatorial Identities

$$= \binom{2015}{10000} + \binom{2014}{1000} + \cdots + \binom{1001}{1000} + \binom{999}{999}$$

$$= \binom{2015}{10000} + \binom{2014}{1000} + \cdots + \binom{1001}{1000} + \binom{1000}{1000}$$

$$= \binom{2016}{1001}$$

It follows that the final result is

$$M = \binom{2016}{1001} \div \binom{2015}{1000} = \frac{2016}{1001} = \boxed{\frac{288}{143}}$$

Done.

5.8 Change the Index

There are dozens, if not hundreds or even thousands of, combinatorial identities. It is challenging to remember them all. Having said this, it is more important to master a handful of core identities and their associated techniques. Many of them have already been covered in this chapter. Additionally, the following changing-of-index identity, together with the Pascal triangle identity *(5.4)* on *page 47*, often serve as important stepping stone in deriving some complex relations.

$$\binom{n}{k} = \frac{k+1}{n+1}\binom{n+1}{k+1} \qquad (5.17)$$

Let's consider a PUMaC problem as an example to show the applications of *(5.17)*.

Example 5.8.1

Given 5 randomly selected distinct positive integers not exceeding 90, what is the expected average value of the fourth largest number?

Chapter 5: Combinatorial Identities

Solution

Let k be the fourth largest number. Then it must satisfy $3 < k < 90$. Meanwhile, for k to be the fourth largest number, there must be three numbers chosen from 1 to $(k-1)$, inclusive, and one number chosen from $(k+1)$ to 90, inclusive. Therefore, there are totally $(90-k)\binom{k-1}{3}$ different ways this scenario can occur. It follows that the expected value is

$$E = \frac{1}{\binom{90}{5}} \underbrace{\sum_{k=1}^{89} k(90-k)\binom{k-1}{3}}_{A} \tag{5.18}$$

where $\binom{90}{5}$ is the number of different ways to choose 5 distinct number from 90 candidates.

By (5.17), we have $k \cdot \binom{k-1}{3} = 4 \cdot \binom{k}{4}$. Therefore

$$A = \sum_{k=1}^{89} k(90-k)\binom{k-1}{3}$$

$$= 4\sum_{k=1}^{89}(90-k)\binom{k}{4}$$

$$= 4\sum_{k=1}^{89}(91-(k+1))\binom{k}{4}$$

$$= 4\cdot 91 \underbrace{\sum_{k=1}^{89}\binom{k}{4}}_{B} - 4\underbrace{\sum_{k=1}^{89}(k+1)\binom{k}{4}}_{C}$$

Part B above can be evaluated using the hockey stick identity

$$B = \sum_{k=1}^{89}\binom{k}{4} = \sum_{k=4}^{89}\binom{k}{4} = \binom{90}{5}$$

Part C can be transformed again by identity $(k+1)\binom{k}{4} = 5\binom{k+1}{5}$

Chapter 5: Combinatorial Identities

before applying the hockey stick identity

$$C = \sum_{k=1}^{89}(k+1)\binom{k}{4} = 5\sum_{k=1}^{89}\binom{k+1}{5} = 5 \cdot \binom{91}{6}$$

Therefore,

$$A = 4 \cdot 91 \cdot \binom{90}{5} - 4 \cdot 5 \cdot \binom{91}{6} = \frac{182}{3}\binom{90}{5}$$

Setting this to (5.18) gives the final answer as $E = \boxed{\dfrac{182}{3}}$

Done.

5.9 Generalized Binomial Theorem

Combinatorial expression is usually only defined for integers n and k satisfying $0 \leq k \leq n$. However the formula itself can hold when n is a negative integer, a real number, or even a complex number.

$$\binom{n}{k} = \frac{n!}{k!(n-k)!} = \frac{n(n-1)(n-2)\cdots(n-k+1)}{1 \cdot 2 \cdots k} \quad (5.19)$$

According to (5.19), $\binom{n}{k} = 0$ when $k > n$. This result still makes intuitive sense because there is no way to select more objects from available candidates. However, when n is a negative, real or complex number, (5.19) has no apparent combinatorial meaning though it can still produce a numerical value. Nevertheless, such extended definitions have some useful advanced applications.

$$\binom{-5}{3} = \frac{(-5) \times (-6) \times (-7)}{1 \times 2 \times 3} = -35 \quad (5.20)$$

and

$$\binom{-2.5}{3} = \frac{(-2.5) \times (-3.5) \times (-4.5)}{1 \times 2 \times 3} = -6.5625$$

Chapter 5: Combinatorial Identities

For positive n, it holds that

$$\binom{-n}{k} = (-1)^k \binom{n+k-1}{k} \tag{5.21}$$

Hence, *(5.20)* can also be computed as

$$\binom{-5}{3} = (-1)^3 \binom{5+3-1}{3} = -35$$

It is worth pointing out that, by convention, $\binom{n}{0} = 1$ regardless of n.

It turns out that the binomial theorem also holds when the exponent is a real or a complex number.

Theorem 5.9.1 Generalized Binomial Theorem

(Generalized Binomial Theorem) Let r be a complex number, then the following relation holds

$$(a+b)^r = \sum_{k=0}^{\infty} \binom{r}{k} a^{r-k} b^k \tag{5.22}$$

Equation 5.22 will become the regular binomial theorem when r is a positive integer. This is because $\binom{n}{k} = 0$ when $n < k$. Proving generalized binomial theorem requires calculus. Understanding its proof is not essential in high school math competitions.

Proof

Let's consider function $f(b) = (a+b)^r$ where a and r are constants. Taking k^{th} derivative on both sides gives

$$\frac{d^k}{db^k} f(b) = r(r-1) \cdots (r-k+1)(a+b)^{r-k}$$

Chapter 5: Combinatorial Identities

This means

$$\frac{d^k}{db^k}f(0) = r(r-1)\cdots(r-k+1)a^{r-k}$$

Applying Taylor expansion centered at 0 gives

$$(a+b)^r = \sum_{k=0}^{\infty} \frac{r(r-1)\cdots(r-k+1)a^{r-k}}{k!}b^k = \sum_{k=0}^{\infty} \binom{r}{k}a^{r-k}b^k$$

QED

5.10 Practice

Practice 1

Simplify the expression

$$S = \binom{2020}{0}^2 + \binom{2020}{1}^2 + \cdots + \binom{2020}{2020}^2$$

(ref: 2721)

Practice 2

Given

$$P(x) = (1 + x + x^2)^{100} = a_0 + a_1 x + \cdots + a_{200} x^{200}$$

Compute the following sums:

- $S_1 = a_0 + a_1 + a_2 + a_3 + \cdots + a_{200}$
- $S_2 = a_0 + a_2 + a_4 + a_6 + \cdots + a_{200}$.

(ref: 3978)

Practice 3

For some particular value of N, when $(a + b + c + d + 1)^N$ is expanded and like terms are combined, the resulting expression contains exactly 1001 terms that include all four variables $a, b, c,$ and d, each to some positive power. What is N?

(ref: 2884 - AMC10)

Chapter 5: Combinatorial Identities

Practice 4

For any positive integer n and prime p, it must hold that

$$(1+n)^p \equiv 1+n^p \pmod{p}$$

Practice 5

Evaluate the value of

$$\sum_{m=0}^{2009} \sum_{n=0}^{m} \binom{2009}{m}\binom{m}{n}$$

(ref: 4323 - PUMaC)

Practice 6

Let $\mathbb{S}$ be a set of integers, $\max(\mathbb{S})$ be the largest element in $\mathbb{S}$, and $|\mathbb{S}|$ be the number of elements in $\mathbb{S}$. Find the number of non-empty set $\mathbb{S} \in \{1, 2, \cdots, 10\}$ satisfying $\max(\mathbb{S}) \leq |\mathbb{S}| + 2$.

(ref: 4327 - PUMaC)

Practice 7

Let m and n be positive integers. Show that

$$\frac{(m+n)!}{(m+n)^{m+n}} < \frac{m!\, n!}{m^m\, n^n}$$

(ref: 4328 - Putnam)

Practice 8

Let n be an even integer. Find the number of ways to select four distinct integers a, b, c, d between 1 and n, inclusive, satisfying $a + c = b + d$. Order of these four numbers does not matter.

(ref: 4417)

Practice 9

Find the sum of all n such that

$$\binom{n}{0} - \binom{n}{1} + \binom{n}{2} - \binom{n}{3} + \cdots + \binom{n}{2018} = 0$$

(ref: 4324)

Practice 10

How many 3×3 matrices of non-negative integers are there such that the sum of every row and every column equals n?

(ref: 4425)

Practice 11

Show that for any positive integer n, the value of $\frac{(n^2)!}{(n!)^{n+1}}$ is an integer.

(ref: 4363)

Chapter 5: Combinatorial Identities

Chapter 6

Generating Function

6.1 Introducing Generating Function

The generating function method is a powerful technique which can be used to solve a variety of problems. In the context of combinatorics, this method models a problem using polynomial operations in such a way that the desired result is the coefficient of a specific term. The idea of utilizing coefficients is already used in *Example 5.3.1* on *page 45* when the coefficient method is discussed.

In this chapter, generating function will be studied and its application in combinatorics will be explored.

> **Definition 6.1.1 Generating Function**
>
> Given a sequence $\{a_0, a, a_2, \cdots\}$, the polynomial
>
> $$f(x) = a_0 + a_1 x + a_2 x^2 + \cdots + a_n x^n + \cdots$$
>
> is called the generating function for this sequence. The number of terms can be finite or infinite.

Chapter 6: Generating Function

For example, the generating function for the sequence

$$\binom{n}{0}, \binom{n}{1}, \binom{n}{2}, \ldots, \binom{n}{n-1}, \binom{n}{n}$$

is $(1+x)^n$ because

$$(1+x)^n = \binom{n}{0} + \binom{n}{1}x + \binom{n}{2}x^2 + \cdots + \binom{n}{n}x^n$$

Using generating function to solve a combinatorial problem usually involves the following steps:

i) Establish a model, i.e. find the generating function.

ii) Transform and simplify the generating function.

iii) Determine the desired coefficient.

Example 5.3.1 on page 45 has already illustrated these steps. Let's review an example from the previous book Counting.

Example 6.1.1

Roll a standard 6-sided die three times. What is the probability that their sum is 9?

Solution

There are totally $6^3 = 216$ possible outcomes. The number of cases when the sum of three rolls is 9 can be determined using the generating function method. The answer is the coefficient of the x^9 term in the expanded form of the following polynomial

$$f(x) = \left(x + x^2 + x^3 + x^4 + x^5 + x^6\right)^3$$

The outside exponent of 3 corresponds to three rolls. The six terms inside the bracket model the six possible outcomes of each roll. When this expression is expanded before like terms are merged, each one corresponds to a possible sum of three rolls. For example, one x^9 term can be created by any of the following combinations:

$$x \cdot x^2 \cdot x^6, \quad x^2 \cdot x^3 \cdot x^4, \quad \cdots$$

The first combination, $x \cdot x^2 \cdot x^6$, indicates the outcomes of these three rolls are 1, 2, and 6, respectively. Similarly, the second combination above indicates the outcomes as 2, 3, and 4, respectively. Therefore, when like terms are merged, their coefficients are simply the numbers of possible combinations. Hence, the coefficient of x^9 is the number of possibilities which have a sum of 9.

Expanding $f(x)$ directly gives the coefficient of x^9 as 25. Therefore the final answer is $\boxed{25/216}$.

Done.

It is worth noting that while direct expanding a polynomial may be convenient in some cases, utilizing generating function prperties and some well-known results can be easier in other scenarios. They will be discussed next.

6.2 Generating Function Properties

Let $f(x) = \sum_{k=0}^{\infty} a_k x^k$ and $g(x) = \sum_{k=0}^{\infty} b_k x^k$, then

i) $f(x) = g(x)$ if and only if $a_k = b_k$ holds for all k

ii) $f(x) \pm g(x) = \sum_{k=0}^{\infty} (a_k \pm b_k) x^k$

iii) $f(x)g(x) = \sum_{k=0}^{\infty} \left(\sum_{j=0}^{k} a_j b_{k-j} \right) x^k$

Chapter 6: Generating Function

iv) The quotient $\frac{f(x)}{g(x)}$ exists if and only if there exists a polynomial $h(x)$ such that $h(x)g(x) = f(x)$

These properties are all self-explanatory. The 3rd property often plays a key role in problem solving because studying coefficients is the focus. Let's review *Example 5.3.1* on *page 45* whose target is to find the coefficient of the term x^r in the expanded form of

$$\left(\binom{m}{0} + x\binom{m}{1} + \cdots x^m\binom{m}{m}\right)\left(\binom{n}{0} + x\binom{n}{1} + \cdots x^n\binom{n}{n}\right)$$

Let

$$f(x) = \sum_{k=0}^{m} \binom{m}{k} x^k, \quad g(x) = \sum_{k=0}^{n} \binom{n}{k} x^k$$

Then applying the 3rd property and noting the last term in the result is x^{m+n} give

$$f(x)g(x) = \sum_{k=0}^{m+n} \left(\sum_{j=0}^{k} \binom{m}{j}\binom{n}{k-j}\right) x^k$$

Setting $k = r$ gives the coefficient of the term x^r as

$$\sum_{j=0}^{r} \binom{m}{j}\binom{n}{r-j}$$

This agrees with the result given in *Example 5.3.1*.

6.3 Useful Conclusions and Techniques

In order to effectively use the generating function method, the ability to find the corresponding polynomial and manipulate it is vital. The conclusions given in this section may be used directly as well-known results and serve as building blocks to solve more complex problems.

Chapter 6: Generating Function

Example 6.3.1

Show that
$$\frac{1}{1-x} = 1 + x + x^2 + x^3 + \cdots \qquad (6.1)$$

Proof

Let $f(x) = 1 - x$. Then, the goal is to find
$$g(x) = a_0 + a_1 x + a_2 x^2 + a_3 x^3 + \cdots$$
which satisfies $f(x)g(x) = 1$. Because
$$\begin{aligned} f(x)g(x) &= (1-x)g(x) \\ &= g(x) - xg(x) \\ &= (a_0 + a_1 x + a_2 x^2 + \cdots) - x(a_0 + a_1 x + a_2 x^2 + \cdots) \\ 1 &= a_0 + (a_1 - a_0)x + (a_2 - a_1)x^2 + (a_3 - a_2)x^3 + \cdots \end{aligned}$$
therefore, it must hold that
$$a_0 = 1, \quad \text{and} \quad a_{i+1} - a_i = 0, \ i = 1, 2, 3, \cdots$$
This means that $a_0 = a_1 = a_2 = a_3 = \cdots = 1$ or
$$g(x) = 1 + x + x^2 + x^3 + \cdots$$

QED

In the context of generating function, it is the coefficients that matter. As such, it is not meaningful to assign x with a specific value (such as $x = 2$) in order to validate or invalidate the relation. That being said, if such an identity is used for something beyond comparing coefficients, then appropriate value must be chosen so that the value of both sides of the identity will equal. For example, (6.1) can hold numerically only if and only if $|x| < 1$.

The theorem below is a generalization of *Equation 6.1*.

Chapter 6: Generating Function

> **Theorem 6.3.1 Generating Function for $\frac{1}{(1-x)^n}$**
>
> $$\frac{1}{(1-x)^n} = \sum_{k=0}^{\infty} \binom{n-1+k}{n-1} x^k \tag{6.2}$$

One way to prove *Theorem 6.3.1* is using mathematical induction. Interested readers can have a try as a practice.

Example 6.3.2

Find the generating function for the sequence (an infinite number of 1s with k leading 0s)

$$\underbrace{0, ,0 , \cdots , 0,}_{k} 1, 1, , 1, 1, 1, \cdots$$

Solution

The conclusion of *Equation 6.1* states that $f(x) = \frac{1}{1-x}$ is the generating function for the sequence

$$1, 1, 1, ,1 , \cdots$$

Then, it can be verified that the answer to this problem is

$$x^k f(x) = \frac{x^k}{1-x}$$

Done.

This example can be summarized into the following theorem.

Theorem 6.3.2 Right Shifting

Let $f(x)$ be the generating function for the sequence

$$a_0, \ a_1, \ a_2, \ a_3, \ \cdots$$

then, the generating function for the right shifted sequence

$$\underbrace{0, \ ,0 \ , \ \cdots \ , \ 0,}_{k} \ a_0, \ a_1, \ a_2, \ a_3, \ a_4, \ \cdots$$

is
$$x^k f(x)$$

Similarly, it can be show that

Theorem 6.3.3 Left Shifting

Let $f(x)$ be the generating function for the sequence

$$a_0, \ a_1, \ a_2, \ a_3, \ \cdots$$

then, the generating function for the right shifted sequence

$$a_1, \ a_2, \ a_3, \ a_4, \ \cdots$$

is
$$\frac{f(x) - a_0}{x}$$

The next theorem is about partial sum.

Chapter 6: Generating Function

Theorem 6.3.4   Partial Sum

Let $f(x)$ be the generating function for $a_0, a_1, a_2, \cdots$. Then the generating function for

$$a_0, \; a_0 + a_1, \; a_0 + a_1 + a_2, \; \cdots$$

is

$$f(x) \sum_{n=0}^{\infty} x^n = f(x) \left(1 + x + x^2 + x^3 + \cdots \right)$$

Proof

The given condition implies

$$f(x) = \sum_{n=0}^{\infty} a_n x^n$$

Then

$$f(x) \sum_{n=0}^{\infty} x^n = \left(\sum_{n=0}^{\infty} a_n x^n \right) \left(\sum_{n=0}^{\infty} x^n \right) = \sum_{n=0}^{\infty} \left(\sum_{k=0}^{n} a_n \right) x^n$$

$$= \sum_{n=0}^{\infty} (a_0 + a_1 + \cdots + a_n) x^n$$

QED

The next example can be directly obtained using *Equation 6.2* above. Meanwhile, it can also be served as a good example of the derivative method.

Example 6.3.3

Show that

$$\frac{1}{(1-x)^2} = 1 + 2x + 3x^2 + 4x^3 + \cdots \qquad (6.3)$$

Proof

Taking derivative with respect to x on *Equation 6.1* on *page 69* leads to the conclusion immediately.

$$QED$$

Theorem 6.3.5 Differentiation

Let $f(x)$ be the generating function for the sequence

$$a_0,\ a_1,\ a_2,\ \cdots$$

Then $f'(x)$ is the generating function for the sequence

$$a_1,\ 2a_2,\ 3a_3,\ \cdots$$

In other words, the effect of taking derivative is to multiply each term by its position index, and then left shift all terms by one. The first term in the original sequence will be eliminated.

6.4 Integer Solution Generalization

The integer solution discussed in *Chapter 4* can be generalized in several different ways. Two of them are

i) The value of at least one variable is bounded.

ii) The coefficient is an integer other than 1.

These two cases can be solved using the generating function method. *Example 6.1.1* on *page 66* falls into the first category because it can be modeled as finding the number of integer solutions to the equation

$$x_1 + x_2 + x_3 = 9$$

Chapter 6: Generating Function

where $1 \leq x_1, x_2, x_3 \leq 6$.

Below is an example for the second case.

Example 6.4.1

Find the number of non-negative integer solutions to the equation
$$x + 2y + 4z = 6$$

Solution

The answer is the coefficient of term x^6 in the expanded form of
$$f(x) = (1 + x + x^2 + x^3 + \cdots)(1 + x^2 + x^4 + \cdots)(1 + x^4 + x^8 + \cdots)$$

This particular problem can be simplified using casework because the contributing term from the third bracket can only be 1 or x^4. Accordingly, the answer is the sum of the coefficients of terms x^6 and x^2, respectively in
$$g(x) = (1 + x + x^2 + x^3 + \cdots)(1 + x^2 + x^4 + \cdots)$$

Because
$$\begin{aligned} g(x) &= \frac{1}{1-x} \cdot \frac{1}{1-x^2} \\ &= \frac{1}{(1-x)^2(1+x)} \\ &= \frac{1}{4} \cdot \left(\frac{1}{1-x} + \frac{2}{(1-x)^2} + \frac{1}{1+x} \right) \end{aligned} \qquad (6.4)$$

Therefore
$$\begin{aligned} 4g(x) &= (1 + x + x^2 + x^3 + x^4 + x^5 + x^6 + \cdots) \\ &\quad + 2 \cdot (1 + 2x + 3x^2 + 4x^3 + 5x^4 + 6x^5 + 7x^6 + \cdots) \\ &\quad + (1 - x + x^2 - x^3 + x^4 - x^5 + x^6 + \cdots) \\ &= 4 + 4x + 8x^2 + \cdots + 16x^6 + \cdots \end{aligned}$$

Chapter 6: Generating Function

It follows that the final result is
$$\frac{1}{4} \cdot (8 + 16) = \boxed{6}$$

Done.

Please note that the decomposition technique used in deriving the step *(6.4)* is discussed in the book Power Calculation.

6.5 More Examples

The next example shows that the generating function method is compatible with traditional less sophisticated methods. Some problems that can be solved using simpler techniques can also be tackled using generating function.

Example 6.5.1

How many different ways are there to form a 6-student team from a pool of 100 candidates?

Solution

Using the basic counting technique, the answer is $\binom{100}{6}$. This result can also be obtained by checking the coefficient of the x^6 term in the expanded form of $(1 + x)^{100}$. The constant $1 = x^0$ means the corresponding student is not selected while the term x represents the case when this student is chosen.

Done.

Here is an example from the PUMac contest.

Chapter 6: Generating Function

Example 6.5.2

For every integer n from 0 to 6, we have 3 identical weights which weight 2^n grams. How many ways are there to form a total weight of 263 grams using only these weights?

Solution

This problem is a typical one which can be solved using the generating function method. The answer is the coefficient of the x^{263} in the expanded form of the following polynomial.

$$f(x) = (1 + x + x^2 + x^3)(1 + x^2 + x^{2\times 2} + x^{2\times 3})$$

$$\cdots$$

$$(1 + x^{2^6} + x^{2^6 \times 2} + x^{2^6 \times 3})$$

$$= \frac{x^4 - 1}{x - 1} \cdot \frac{x^8 - 1}{x^2 - 1} \cdots \frac{x^{256} - 1}{x^{64} - 1}$$

$$= \frac{x^{128} - 1}{x^2 - 1} \cdot \frac{x^{256} - 1}{x - 1}$$

$$= \left(1 + x^2 + x^4 \cdots + x^{126}\right)\left(1 + x + x^2 + \cdots + x^{255}\right)$$

In order to obtain one x^{263} term, we must take one term from the first bracket whose exponent is at least 8. There are a total of 60 such terms. For each of these terms, there is one unique matching term in the second bracket to make their product x^{263}. Hence, the desired result is $\boxed{60}$.

Done.

6.6 Practice

Practice 1

How many different 5-digit numbers can be formed using 1, 2, 3, and 4 that satisfy the following conditions:

- the digit 1 must appear either 2 or 3 times,
- the digit 2 must appear even times,
- the digit 3 must appear odd times, and
- the digit 4 has no restriction

(ref: 2536)

Practice 2

Find the number of integer solutions to the equation $a+b+c=6$ where $-1 \leq a < 2$ and $1 \leq b,\ c \leq 4$.

(ref: 4360)

Practice 3

How many different ways are there to make a payment of n dollars using any number of \$1 and \$2 bills?

(ref: 4329)

Practice 4

How many 4-digit integers are there whose sum of all digits equals 12?

(ref: 4330)

Chapter 6: Generating Function

Practice 5

There are 30 identical souvenirs to be distributed among 50 students. Each student may receive more than one souvenir. How many different distribution plans are there?

(ref: 4359)

Practice 6

A deck of poker has three different colors each of which contains 10 cards numbered from 1 to 10, respectively. In addition, there are two jokers both of which are numbered as 0. A card with number k is valued as 2^k points. How many different ways are there to draw several cards from this deck so that their total value equals 2004?

(ref: 4333 - China)

Practice 7

Let n be a positive integer. Find the number a_n of polynomials $f(x)$ with coefficients in $\{0, 1, 2, 3\}$ such that $f(2) = n$.

(ref: 4335)

Practice 8

Let a_0, a_1, a_2, $\cdots$ be an increasing sequence of non-negative integers such that every non-negative integer can be expressed uniquely in the form of $(a_i + 2a_j + 4a_k)$ where i, j, and k are not necessarily distinct. Determine a_{1998}.

(ref: 4336 - IMO)

Chapter 6: Generating Function

Practice 9

Show that

$$\sqrt{1+x} = 1 + \sum_{n=1}^{\infty} \frac{(-1)^{n-1}}{n \cdot 2^{2n-1}} \binom{2n-2}{n-1} x^n$$

(ref: 4314)

Practice 10

Complete the last two practices in *Chapter 3* which is to solve the recursion

$$a_n = \sum_{k=0}^{n-1} a_k a_{n-k-1} = a_0 a_{n-1} + a_1 a_{n-2} + \cdots + a_{n-1} a_0$$

where $a_0 = a_1 = 1$.

(ref: 4511)

Chapter 6: Generating Function

Appendices

Appendix A

Solutions

Chapter A: Solutions

A.1 *Chapter 1*

This section is intentionally left blank.

So section numbers of solutions and practices can match.

A.2 Chapter 2

Practice 1

Explain the identity $\binom{n}{k} = \binom{n}{n-k}$ using bijection.

Every selection where k items are chosen means the remaining $(n-k)$ items are not selected. On the other hand, every selection where $(n-k)$ items are not chosen means the remaining k items are selected. Hence, choosing k items has a bijective relation with not selecting the remaining $(n-k)$ items. This means

$$\binom{n}{k} = \binom{n}{n-k}$$

Practice 2

Given a convex n-polygon, what is the max number of intersection points its diagonals can form? (Vertices do not count.)

(ref: 4364)

A maximum can be achieved if no three diagonals meet at the same point. In this case, any intersection point has a bijective relation with four distinct vertices. This is because assuming there are four vertices A, B, C, and D in that order but not necessarily next to each other, then diagonals AC and BD will create an intersection point. On the other hand, any intersection point created by diagonals AC and BD must have their ending points selected from four distinct vertices of the n-polygon.

Then, it follows that the answer is $\boxed{\binom{n}{4}}$.

Chapter A: Solutions

Practice 3

How many fractions in simplest form are there between 0 and 1 such that the products of their denominators and numerators equal 20!?

(ref: 2717)

We note that 20! has 8 different prime factors. Each prime factor, together with its exponent in the prime factorization of 20! can be part of either the denominator or the numerator, but not both, in order to make the result irreducible. Hence, there are 2^8 different combinations. Meanwhile, every case in which a prime factor goes to denominator must correspond to a case in which this prime factor goes to the numerator. These two cases form a bijection. In every such pair, only one resulting fraction will be less that 1. Therefore, we conclude the answer is

$$2^7 = \boxed{128}$$

Practice 4

A child builds towers using identically shaped cubes of different colors. How many different towers with a height of 8 cubes can the child build with 2 red cubes, 3 blue cubes, and 4 green cubes? (One cube will be left out.)

(ref: 4385 - AMC10)

The answer is $\boxed{1260}$. This is because arrangements of 8 blocks have a bijective relation with those of 9 blocks by simply removing the last block. Meanwhile, the number of arrangements of 9 blocks can be computed as

$$\frac{9!}{2! \cdot 3! \cdot 4!} = \boxed{1260}$$

Chapter A: Solutions

Practice 5

There are $n \geq 6$ points on a circle, every two points are connected by a line segment. No three diagonals are concurrent. How many triangles are created by these sides and diagonals?

(ref: 4368 - China)

Let's call the n points on the circle as circle-points and those intersection points as inner-points. Then triangles formed by the circle-points and inner-points can be grouped into the following four categories:

- $\mathbb{A}$: all vertices are circle-points
- $\mathbb{B}$: two vertices are circle-points and one is inner-point
- $\mathbb{C}$: one vertex is circle-point and two are inner-points
- $\mathbb{D}$: all vertices are inner-points

Every triangle in $\mathbb{A}$ has a bijective relation with three distinct points on the circle. Hence, $|\mathbb{A}| = \binom{n}{3}$ where $|\mathbb{S}|$ denotes the number of elements in set $\mathbb{S}$.

For $\mathbb{B}$, let's consider any inner-point O as shown in the first diagram below. An inner point O determines four triangles in $\mathbb{B}$. Meanwhile, O has a bijective relation with four circle-points. This means that these five points have a bijective relation with four triangles in $\mathbb{B}$: $\triangle A_1 A_2 O$, $\triangle A_2 A_3 O$, $\triangle A_3 A_4 O$ and $\triangle A_4 A_1 O$. Hence, $|\mathbb{B}| = 4\binom{n}{4}$.

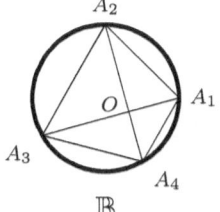
$\mathbb{B}$

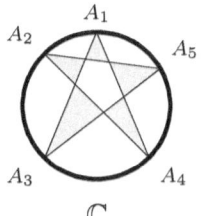
$\mathbb{C}$

Chapter A: Solutions

For $\mathbb{C}$, as shown in the right graph above, every 5 circle-points have a bijective relation with 5 inner-points. All these 10 points have a bijective relation with 5 triangles in $\mathbb{C}$. Hence, $|\mathbb{C}| = 5\binom{n}{5}$.

For the last case, there are $|\mathbb{D}| = \binom{n}{6}$ such triangles. This is because given any six points $A_1, A_2, A_3, A_4, A_5, A_6$ in that order but not necessarily next to each other, the intersection points created by segments A_1A_4, A_2A_5 and A_3A_6 has a one-to-one mapping with a triangle in $\mathbb{D}$.

Therefore, the final answer is

$$\boxed{\binom{n}{3} + 4\binom{n}{4} + 5\binom{n}{5} + \binom{n}{6}}$$

Practice 6

John walks from point A to C while Mary goes from point B to D. Both of them will move along the grid, either right or up, so they take shortest routes. How many different possibilities are there such that their routes do not intersect?

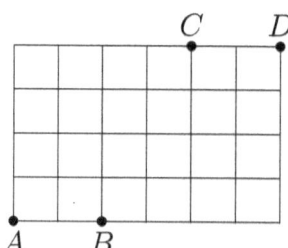

(ref: 3159)

This problem can be solved in several different ways. One of them is to count the opposite cases using bijection.

Without any restriction, there are $\binom{8}{4}$ different paths for John and the same number of paths for Mary. Hence, there are total $\binom{8}{4}^2$ different combinations.

Next, instead of counting the number of paths that do not intersect, let's count the number of intersecting paths. Assuming two such paths first insect at point X. Then the original paths (AXC, BXD) will have a bijection with (AXD, BXC). (Imagine that John and Mary switch their destinations at the first intersection point.)

Because there are $\binom{10}{4}$ paths from A to D, and $\binom{6}{2}$ ways from B to C, so the number of intersecting paths is $\binom{10}{4}\binom{6}{2}$.

It follows that the final answer is

$$\binom{8}{4}^2 - \binom{10}{4}\binom{6}{2} = \boxed{1750}$$

Practice 7

Let N be the number of non-decrease sequences of length n and each element is a non-negative integer not exceeding d. Show that N equals the number of non-negative solutions to the following equation:

$$x_0 + x_1 + \cdots + x_d = n$$

Let x_i $(0 \le i \le d)$ in the given equation represent the number of integer i appeared in the sequence. Then, each x_i is a non-negative value. Meanwhile, because their sum is n, therefore there are totally n numbers in the sequence.

For any solution to the equation, or equivalently a collection of n non-negative integers not exceeding d, there is only one way to form a non-decreasing sequence. Hence, the number of solutions to the given equation equals the number of qualified non-decreasing

Chapter A: Solutions

sequences.

Practice 8

A partition of a positive integer n is to write n as a sum of some positive integers. Let k be a positive integer. Show that the number of partitions of n with exactly k parts equals the number of partitions of n whose largest part is exactly k.

(ref: 4352)

This claim can be explained by visualization. Consider a partition of n as
$$n = a_1 + a_2 + \cdots + a_k$$
where $a_1 \geq a_2 \geq \cdots \geq a_k$. This partition can be represented by an array of dots. The left part of the following diagram illustrates an example of $10 = 5 + 3 + 1 + 1$. Then, this diagram can flipped along its main diagonal to create a paired partition whose largest part equals the number of partitions in the original partition.

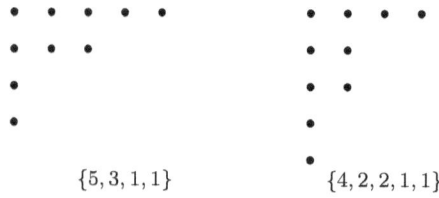

Meanwhile, the right side of the diagram can also be mapped back to the left side. Hence, they have a one-to-one mapping relationship. This means that the claim holds.

Chapter A: Solutions

Practice 9

Show that the number of partitions of a positive integer n into distinct parts is equal to the number of partitions of n where all parts are odd integers.

(ref: 4353)

Our goal is to establish a bijection between these two ways of partitions.

Given a partition of distinct parts, write every part in the form of $a \cdot 2^b$ where a is an odd integer and b is a non-negative integer. Then, we can split every part into 2^b of a. All parts will be odd integers. For example, given a partition of $15 = 10 + 4 + 1$, the following transformation can be performed to obtain an all-odd partition:

$$15 = 10 + 4 + 1$$
$$= 5 \cdot 2^1 + 1 \cdot 2^2 + 1 \cdot 2^0$$
$$= (5 + 5) + (1 + 1 + 1 + 1) + (1)$$
$$= 5 + 5 + 1 + 1 + 1 + 1 + 1$$

On the other direction, we first group equal parts. Assuming there are k parts of a and $k = 2^{k_1} + 2^{k+2} + \cdots + 2^{k_p}$ where $k_1 > k_2 > \cdots > k_p$. Then the sum of these parts can be written as

$$k \cdot a = a \cdot 2^{k_1} + a \cdot 2^{k_2} + \cdots + a \cdot 2^{k_p}$$

Hence, it is possible to convert these parts to $a \cdot 2^{k_1}$, $a \cdot 2^{k_1}$, $\cdots$, $a \cdot 2^{k_p}$ which are all distinct. Different groups of such equal parts will result in different terms because $a \cdot 2^m$ will never equal $b \cdot 2^n$ if both a and b are different odd integers. For example, the partition $15 = 5 + 5 + 1 + 1 + 1 + 1 + 1$ can be transformed back to $10 + 4 + 1$ via

$$15 = 5 + 5 + 1 + 1 + 1 + 1 + 1$$
$$= 5 \cdot 2 + 1 \cdot 5$$
$$= 5 \cdot (2^1) + 1 \cdot (2^2 + 2^0)$$

$$= 5 \cdot 2^1 + 1 \cdot 2^2 + 1 \cdot 2^0$$
$$= 10 + 4 + 1$$

Hence, there is indeed a bijection between these two cases. This means their counts are equal.

Practice 10

Let positive integers n and k satisfy $n \geq 2k$. How many k-sided convex polygons are there whose vertices are those of an n-sided convex polygon and edges are diagonals of the same n-polygon.

(ref: 4362)

Label the vertices of this n-polygon as $1, 2, \cdots, n$ and the vertices of a qualified k-polygon be $A_1, A_2, \cdots, A_k$. Then there are two possible cases:

- $A_1 = 1$, $3 \leq A_2 < A_3 < \cdots < A_k \leq (n-1)$, and $A_{i+1} - A_i \geq 2$ for all $2 \leq i \leq (k-1)$

- $2 \leq A_1 < A_2 < \cdots < A_k \leq n$, and $A_{i+1} - A_i \geq 2$ for all $1 \leq i \leq (k-1)$

In the first scenario above, let's transform $\mathbb{A} = \{A_2, A_3, \cdots, A_k\}$ to $\mathbb{A}' = \{A_2 - 2, A_3 - 3, \cdots, A_k - k\}$. This transformation is bijective because every element in $\mathbb{A}$ has a corrsponding one in $\mathbb{A}'$, and vice vesa. Meanwhile, elements in $\mathbb{A}'$ satisfy

$$1 \leq A_2 - 2 < A_3 - 3 < \cdots < A_k - k \leq n - k - 1$$

A selection of $\mathbb{A}'$ is to choose $(k-1)$ distinct numbers from $\{1, 2, \cdots, n-k-1\}$ which has $\binom{n-k-1}{k-1}$ ways.

In the latter case, let's transform $\mathbb{A} = \{A_1, A_2, A_3, \cdots, A_k\}$ to $\mathbb{A}' = \{A_1 - 1, A_2 - 2, A_3 - 3, \cdots, A_k - k\}$. This transformation is also bijective. At the same time, elements in $\mathbb{A}'$ satisfy

$$1 \leq A_1 - 1 < A_2 - 2 < A_3 - 3 < \cdots < A_k - k \leq n - k$$

and form a selection of k distinct numbers from $\{1, 2, \cdots, n-k\}$. There are $\binom{n-k}{k}$ ways.

Hence, the final answer is

$$\binom{n-k-1}{k-1} + \binom{n-k}{k} = \boxed{\frac{n}{k}\binom{n-k-1}{k-1}}$$

Practice 11

A tree is a structure which grows from a single root node. By convention, the root node is usually placed at the top. Then, new nodes, referred as children nodes, can be attached to an existing node, referred as parent node, by edges. The root node has no parent and all other nodes have exactly one parent. A node may have any number of children nodes, including no child at all. No looping chain of nodes is permitted in this structure. For example, there are exactly 5 different types of trees with 4 nodes. They are shown below. The goal is to find the number of different types of tree with n nodes where n is a positive integer greater than 1.

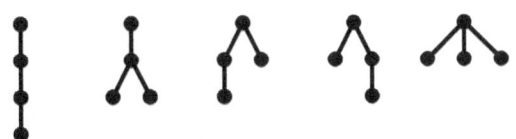

(ref: 4357)

The answer is $\boxed{\dfrac{1}{n}\binom{2(n-1)}{n-1}}$. Or equivalently, there are $\dfrac{1}{n+1}\binom{2n}{n}$ different types for a tree of $(n+1)$ nodes.

A bijection can be established between this problem and *Example 2.2.2* on *page 6* using a deep first traversal. The deep first traversal is a recursive visiting algorithm that defines which node to visit next

Chapter A: Solutions

after the current node:

- If the current node dose not have a child node or all its children have been visited, then move back to its parent node

- Otherwise, visit the first child node which has not been visited

The visiting sequence starts and terminates at the root node. For example, in the following tree, the visiting sequence will be $0-1-2-3-2-1-4-1-0$.

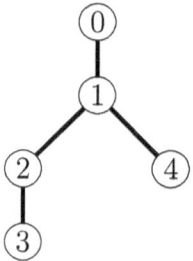

Now, let's map such a traversal to a properly matched parentheses sequence in *Example 2.2.2* on *page 6*. A left parenthesis is written when a node, other than the root node, appears the first time in the visiting sequence and a right parenthesis is written when this node appear the last time in the sequence. When a node appears only once in the traversal (i.e. this node has no child), both a left parenthesis and a right one are written. For example, the traversal shown in the diagram above corresponds to the following parentheses sequence:

$$((())())$$

Therefore, a tree with $(n+1)$ nodes corresponds to a well matched sequence of n parentheses because the root node does not match any parenthesis. Then the result follows.

Chapter A: Solutions

Practice 12

Find the number of parallelograms in the following equilateral triangle of side length n which is made of some smaller unit equilateral triangles.

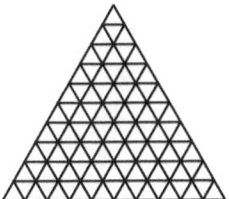

(ref: 4354)

There are three possible shapes of parallelograms in this diagram which are shown below.

Each type has two adjacent sides paralleling to the two edges of the big equilateral triangle. Therefore, when the whole picture rotates $\pm 60°$, one type of parallelograms will become another type. This means that the three sets of parallelograms are rotationally symmetric. As a result, it is sufficient to just count one type of parallelograms. The final result is three times of that number.

Without loss of generality, let's count the middle type of parallelograms above. For this, we can extend the bottom of this triangle by one row as shown below.

Chapter A: Solutions

Extending the four edges of a parallelogram to meet the newly added bottom line will lead to four distinct meeting points. Conversely, any four distinct points on the bottom edge will map to a qualified parallelogram. This means that there is a bijection between the set of all such parallelograms and the set of collections of 4 distinct points.

Because there are $\binom{n+1}{4}$ sets containing 4 different points on the bottom line, therefore the final result is

$$\boxed{3\binom{n+1}{4}}$$

Practice 13

Let $\mathbb{S} = \{1, 2, \cdots, 1000\}$ and $\mathbb{A}$ be a subset of $\mathbb{S}$. If the number of elements in $\mathbb{A}$ is 201 and their sum is a multiple of 5, then $\mathbb{A}$ is called *good*. How many good $\mathbb{A}$ are there?

(ref: 4367)

Dividing all subsets of $\mathbb{S}$ containing 201 elements into 5 groups: $\mathbb{S}_0$, $\mathbb{S}_1$, $\mathbb{S}_2$, $\mathbb{S}_3$, and $\mathbb{S}_4$ according to the sum of their elements modulo 5. For example, all subsets having 201 elements whose sums are multiples of 5 are in $\mathbb{S}_0$, whose sums are multiples of 5 plus 1 are in $\mathbb{S}_1$, and so on.

Now, we claim that $\mathbb{S}_0$ and any of $\mathbb{S}_k$, $k = 1, 2, 3, 4$, have a bijective relation. This is because for every $a_i \in \mathbb{S}_0$, there is a unique element $b_i \in \mathbb{S}_k$ satisfying $b_i = a_i + k \pmod{1000}$.

It follows that counts of all these five groups are equal. Meanwhile, the total counts of these five groups is $\binom{1000}{201}$ because this value is equivalent to choosing 201 elements from a 1000 choices in $\mathbb{S}$. Therefore, the number of good $\mathbb{A}$, i.e., $\mathbb{S}_0$, is

$$\frac{1}{5}\binom{1000}{201}$$

Practice 14

Assuming positive integer n satisfies $n \equiv 1 \pmod 4$ and $n > 1$. Let $\mathbb{P} = \{a_1, a_2, \cdots, a_n\}$ be a permutation of $\{1, 2, \cdots, n\}$. If k_p denotes the largest index k associated with $\mathbb{P}$ such that the following inequality holds

$$a_1 + a_2 + \cdots + a_k < a_{k+1} + a_{k+2} + \cdots + a_n$$

Find the sum of k_p for all possible $\mathbb{P}$.

(ref: 4365)

The given condition implies

$$a_1 + a_2 + \cdots + a_{k_p} + a_{k_p+1} \geq a_{k_p+2} + a_{k_p+3} + \cdots + a_n$$

Let's show that the equality cannot hold in the above relation. This is because if

$$a_1 + a_2 + \cdots + a_{k_p} + a_{k_p+1} = a_{k_p+2} + a_{k_p+3} + \cdots + a_n$$

Then

$$2(a_1 + a_2 + \cdots + a_{k_p+1}) = a_1 + a_2 + \cdots + a_n$$
$$= 1 + 2 + 3 + \cdots + n$$

$$= \frac{n(n+1)}{2}$$

Let $n = 4m+1$ where m is a positive integer. Then the right side of the above relation equals $(4m+1)(2m+1)$ which is odd. However, the left side of the above relation is even. Therefore, it is impossible. This means

$$a_1 + a_2 + \cdots + a_{k_p} + a_{k_p+1} > a_{k_p+2} + a_{k_p+3} + \cdots + a_n$$

It follows for any $\mathbb{P} = \{a_1, a_2, \cdots, a_n\}$ with

$$a_1 + a_2 + \cdots + a_{k_p} < a_{k_p+1} + a_{k_p+2} + \cdots + a_n$$

its reversed permutation $\mathbb{P}' = \{a_n, a_{n-1}, \cdots, a_1\}$ must satisfy

$$\underbrace{a_n + \cdots + a_{k_p+2}}_{n-k_p-1} < a_{k_p+1} + a_{k_p} + \cdots + a_1$$

and

$$a_n + \cdots + a_{k_p+2} + a_{k_p+1} > a_{k_p} + \cdots + a_1$$

In other words, the largest indexes of $\mathbb{P}$ and $\mathbb{P}'$ satisfying the given requirement must be k_p and $(n-k_p-1)$ whose sum is $(n-1)$. There are totally $n!$ different permutation. Hence, the answer is

$$\boxed{\frac{(n-1)(n!)}{2}}$$

Practice 15

Let S_n be the number of non-congruent triangles whose sides' lengths are all integers and circumferences equals n. Show that

$$S_{2n-1} - S_{2n} = \left\lfloor \frac{n}{6} \right\rfloor \quad \text{or} \quad \left\lfloor \frac{n}{6} \right\rfloor + 1$$

where $\lfloor x \rfloor$ returns the largest integer not exceeding the real number x.

(ref: 4424)

Chapter A: Solutions

Let $\mathbb{A}_n$ be the set containing all such triangles. Without loss of generality, let the three sides of such a triangle be positive integers a, b, and c satisfying

$$a \geq b \geq c > a - b \quad \text{and} \quad a + b + c = n$$

When $n = 2k$ is even, then it must hold that

$$b \leq a \leq k - 1, \quad \text{and} \quad c \geq 2$$

Define mapping f as

$$(a,\ b,\ c) \Rightarrow (a,\ b,\ c-1)$$

Then f is a mapping from $\mathbb{A}_{2k}$ to $\mathbb{A}_{2k-1}$, or equivalently, from $\mathbb{A}_{2n}$ to $\mathbb{A}_{2n-1}$.

It can be shown that f is an injection, but not a bijection as every element in $\mathbb{A}_{2n}$ has a corresponding one in $\mathbb{A}_{2n-1}$. However, elements in the form of $(a,\ b,\ b) \in \mathbb{A}_{2n-1}$ do not have mappings in $\mathbb{A}_{2n}$. This is because an element $(a,\ b+1,\ b) \in \mathbb{A}_{2n}$ will be mapped to $(a,\ b+1,\ b-1) \in \mathbb{A}_{2n-1}$, not $(a,\ b,\ b)$.

Let

$$\mathbb{A}_{2n} - \mathbb{A}_{2n-1} = \mathbb{B}$$

where $\mathbb{B}$ contains all triangles in the form of $(a,\ b,\ b)$ satisfying

$$a + 2b = 2n - 1 \quad \text{and} \quad b \leq a \leq 2b - 1$$

It follows that $S_{2n} - S_{2n-1}$ equals the number of elements in $\mathbb{B}$, i.e the number of triangles meeting the two conditions above. Solving these two inequalities yields

$$\frac{2n-1}{3} \geq b \geq \frac{n}{2}$$

The count of qualifying b will then be

$$x = \left\lfloor \frac{2n-1}{3} \right\rfloor - \left\lfloor \frac{n}{2} \right\rfloor + 1$$

Chapter A: Solutions

Because when b is determined, a will be determined too (as $a+2b = 2n-1$), therefore the number of elements in $\mathbb{B}$ is also x. Analyzing the modular properties of n will give

$$x = \begin{cases} \left\lfloor \frac{n}{6} \right\rfloor & , \ n \equiv 0, \ 1, \ 3 \pmod 6 \\ \left\lfloor \frac{n}{6} \right\rfloor + 1 & , \ n \equiv 2, \ 4, \ 5 \pmod 6 \end{cases}$$

Hence, the claim holds.

Practice 16

Let $x_i \in \{+1, -1\}$, $i = 1, 2, \cdots, 2n$. If their sum equals 0 and the following inequality holds for any positive integer k satisfying $1 \leq k < 2n$:

$$x_1 + x_2 + \cdots + x_k \geq 0$$

Find the number of possible ordered sequence $\{x_1, x_2, \cdots, x_{2n}\}$.

(ref: 4477)

This problem has a bijection with the one described in *Example 2.2.1 on page 5*.

Every path in that example has a one-to-one mapping with $\{x_1, x_2, \cdots, x_{2n}\}$ when a value of 1 is equivalent to a right-step and a value of -1 is equivalent to an up-step. Therefore, the answer is

$$\boxed{\frac{1}{n+1}\binom{2n}{n}}$$

A.3 Chapter 3

Practice 1

How many different ways are there to cover a 1×10 grid with some 1×1 and 1×2 pieces without overlapping?

(ref: 2495)

Let F_n be the number of ways to cover a $1 \times n$ grid. Then we have $F_1 = 1$ and $F_2 = 2$. For $n \geq 3$, the following recursion holds

$$F_n = F_{n-1} + F_{n-2}$$

depending on the size of the last piece placed:

i) If it is 1×1, then all the previous $1 \times (n-1)$ grids are covered. There are F_{n-1} ways.

ii) If it is 1×2, then all the previous $1 \times (n-2)$ grids are covered. There are F_{n-2} ways.

It follows that $\{F_n\}$ is a Fibonacci sequence. We can either solve this sequence or simply compute the next 8 values starting from F_3 manually. The answer is $F_{10} = \boxed{89}$.

Practice 2

(Hanoi Tower) There are 3 identical rods labeled as A, B, C; and n disks of different sizes which can be slide onto any of these three rods. Initially, the n disks are stacked in ascending order of their sizes on A. What is the minimal number of moves in order to transfer all the disks to B providing that each move can only transfer one disk to another rod's topmost position and at no time, a bigger disk can be placed on top of a smaller one.

(ref: 4478)

Chapter A: Solutions

Let m_n be the required number of moves. Then, it will need a minimal of m_{n-1} moves to transfer the top $(n-1)$ disks from one rod to another. It follows that

i) Transferring top $(n-1)$ disks from A to C requires a minimal of m_{n-1} moves.

ii) Transferring the biggest disk from A to B requires 1 move.

iii) Transferring all the $(n-1)$ disks from C to B requires a minimal of m_{n-1} moves.

Hence, we have

$$m_n = m_{n-1} + 1 + m_{n-1} = 2m_{n-1} + 1$$

with the initial condition $m_1 = 1$. There are several ways to solve this recursion. One is to transform the relation above to

$$(m_n + 1) = 2(m_{n-1} + 1)$$

Then the sequence $\{m_n + 1\}$ is a geometric with initial value of 2 and common ratio 2. Hence

$$m_n + 1 = 2^n \implies m_n = \boxed{2^n - 1}$$

Practice 3

Call a permutation $a_1, a_2, \ldots, a_n$ of the integers $1, 2, \ldots, n$ quasi-increasing if $a_k \leq a_{k+1} + 2$ for each $1 \leq k \leq n - 1$. For example, 53421 and 14253 are quasi-increasing permutations of the integers 1, 2, 3, 4, 5, but 45123 is not. Find the number of quasi-increasing permutations of the integers 1, 2, ..., 7.

(ref: 77 - AIME)

Let F_n be the count of quasi-increasing permutation of the integer $1, 2, \cdots, n$. Then we have $F_1 = 1, F_2 = 2$.

Now consider adding number n to a quasi-increasing sequence from 1 to $(n-1)$. There are only three spots where n can be placed: before $(n-1)$, before $(n-2)$, or at the end. Therefore, we have the recursion when $n > 2$ as

$$F_n = 3 \times F_{n-1} \implies F_n = 2 \times 3^{n-2} \implies F_7 = \boxed{486}$$

Practice 4

In the Banana Country, only Mr Decent always tells the truth and only Mr Joke always tells lies. Everyone else has a probability of p to tell a lie. One day, Mr Decent has decided to run for the President and told his decision to the first person who in turn told this to the second person. The second person then told this to the third person, and so on, till the nth person who told this news to Mr Joke. No one has been told this news twice in this process. Finally, Mr Joke announced Mr Decent's decision to everyone. What is the probability that Mr Joke's statement agrees with Mr Decent's intention?

(ref: 4418)

Let A_n be the probability that Mr Joke's statement agrees with Mr Decent's intention, and $D_n = 1 - A_n$ be the probability that they disagree.

Mr Joke will give the correct statement if and only if the nth person tells him an incorrect statement. This probability depends on the statement of the $(n-1)^{th}$ person which all also influences the value of A_{n-1} and D_{n-1}. The probability of the $(n-1)^{th}$ person to give the correct information equals D_{n-1} and that of a wrong statement is A_{n-1}. Therefore, it should hold that

$$A_n = pD_{n-1} + (1-p)A_{n-1} = p + (1-2p)A_{n-1}$$

This recursion can be rewritten as

$$A_n - A_{n-1} = (1-2p)(A_{n-1} - A_{n-2})$$

Chapter A: Solutions

Meanwhile, when $n = 0$, i.e. Mr Decent shared his decision with Mr Joke directly, $A_0 = 0$. When $n = 1$, $A_1 = p$. Therefore

$$A_n - A_{n-1} = p(1-2p)^{n-1} \implies A_{n-1} = A_n - p(1-2p)^{n-1}$$

Setting this back to the original recursion yields

$$\boxed{A_n = \frac{1 - (1-2p)^n}{2}}$$

Practice 5

How many length ten strings consisting of only As and Bs contain neither "BAB" nor "BBB" as a substring?

(ref: 2046 - Exeter)

This is equivalent to counting the cases where there are no consecutive Bs on every other position. In order to solve this, let's construct two substrings: one consists of the five letters at the odd positions of the original string. The other consists of the five letters at the even positions. We are going to count the cases which there are no consecutive Bs in each of these two substrings.

By the conclusion of *Example 3.2.1* on *page 17*, the answer is $F_5 = 13$. Therefore, the final answer is

$$13 \times 13 = \boxed{169}$$

because these two substrings are indepedent.

Practice 6

Dividing a circle into $n \geq 2$ sectors and coloring these sectors using $m \geq 2$ different colors. If no adjacent sectors can be colored the same, how many different color schemes are there?

(ref: 4407)

Chapter A: Solutions

Let the number of color schemes for n sectors be a_n.

When $n = 2$, one sector has m choices and the other has $(m-1)$ choices. Hence, $a_2 = m(m-1)$.

Next let's show
$$a_n + a_{n-1} = m(m-1)^{n-1}$$
for $n \geq 2$. Let these n sectors be $S_1, S_2, \cdots, S_n$. Then, S_1 has m different colors to choose and S_2 has $(m-1)$ choices. Meanwhile, S_3 has $(m-1)$ colors available to it which are different from S_2. Generally speaking, S_k always have $(m-1)$ options to pick.

Hence, n sectors have $m(m-1)^{n-1}$ such coloring schemes. However, this scheme cannot guarantee S_n and S_1 have different colors. Among the $m(m-1)^{n-1}$ possibilities, there are a_n cases when S_1 and S_n are colored differently by definition. The rest are those situations when S_1 and S_n have the same color. In this case, it must be true that S_1 and S_{n-1} are colored differently. Hence, removing sector S_n will construct a circle of $(n-1)$ sectors with no adjacent sections having the same color. There are a_{n-1} different such $(n-1)$ sectors. In conclusion, we find
$$a_n + a_{n-1} = m(m-1)^{n-1}, \qquad a_2 = m(m-1)$$

In order to solve a_n from this recursion, let $b_n = \frac{a_n}{(m-1)^n}$. Then
$$b_n + \frac{b_{n-1}}{(m-1)} = \frac{m}{m-1} \Longrightarrow (b_n - 1) = -\frac{1}{m-1}(b_{n-1} - 1)$$

This means that $\{b_n - 1\}$ is a geometric sequence whose common ratio is $-\frac{1}{m-1}$ and initial term
$$b_2 = \frac{a_2}{(m-1)^2} - 1 = \frac{m(m-1)}{(m-1)^2} - 1 = \frac{1}{m-1}$$

So
$$b_n = \frac{1}{m-1}\left(-\frac{1}{m-1}\right)^{n-2} = \frac{(-1)^n}{(m-1)^{n-1}}$$

Chapter A: Solutions

It follows that the answer is

$$a_n = b_n(m-1)^n = \boxed{(m-1)^n + (-1)^n(m-1)}$$

Practice 7

There are $2^{10} = 1024$ possible 10-letter strings in which each letter is either an A or a B. Find the number of such strings that do not have more than 3 adjacent letters that are identical.

(ref: 79 - AIME)

Call a string satisfying the stated constraints a good string and let a_n and b_n be the number of good strings of length n which start with letter A and B, respectively.

For $n = 1, 2,$ and 3, we have

$$a_1 = b_1 = 1, \; a_2 = b_2, \; a_3 = b_3 = 4$$

When $n \geq 4$, we can use a similar reasoning as the one used in *Example 3.2.1* on *page 17* to establish the recursion. A good string starting with A must start with AB, AAB or $AAAB$. The letter after the first appeard B has no restriction. This means that, after having removed leading As, a good string starting with A will become a good string starting with B. For example, a good string of length n starting with AB will correspond a good string of length $(n-1)$ starting with B whose count is b_{n-1}. Therefore, we shall have

$$a_n = b_{n-1} + b_{n-2} + b_{n-3}$$

Meanwhile, by the principle of symmetry, it must hold that $a_k = b_k$ for all $k = 1, 2, \cdots, n$. Consequently, we can conclude that

$$a_n = b_{n-1} + b_{n-2} + b_{n-3} \implies a_n = a_{n-1} + a_{n-2} + a_{n-3}$$

Chapter A: Solutions

Using the initial values computed early yields

$$\begin{aligned}
a_4 &= 4 + 2 + 1 & &= 7 \\
a_5 &= 7 + 4 + 2 & &= 13 \\
a_6 &= 13 + 7 + 4 & &= 24 \\
a_7 &= 24 + 13 + 7 & &= 44 \\
a_8 &= 44 + 24 + 13 & &= 81 \\
a_9 &= 81 + 44 + 24 & &= 149 \\
a_{10} &= 149 + 81 + 44 & &= 274 \\
b_{10} &= a_{10}
\end{aligned}$$

Hence, the final answer is $a_{10} + b_{10} = \boxed{548}$.

Practice 8

Let $\mathbb{S} = \{a_1, a_2, \cdots, a_n\}$ be a permutation of $\{1, 2, \cdots, n\}$ which satisfies the condition that for every a_i, ($i = 1, 2, \cdots, n$), there exists an a_j where $i < j \le n$ such that $a_j = a_i + 1$ or $a_j = a_i - 1$. Find the number of such $\mathbb{S}$.

(ref: 4415)

Let the desired count be S_n.

When $n = 1$, the answer is $S_1 = 1$. When $n = 2$, the answer is $S_2 = 2$.

When $n > 2$, let's consider the value of a_1. If $a_1 = 1$ or $a_1 = n$, then the remaining $(n-1)$ elements have S_{n-1} ways all of which satisfy the condition when prefixed with a_1.

If $a_1 = k$ which is neither 1 or n, then $(k+1)$ must be placed after a_1 because there is no element before a_1. Meanwhile, $(k+2)$ must be placed after $(k+1)$ because $(k-1)$ is already placed before k. Repeating this process will lead to the conclusion that $a_n > a_1 = k$.

At the same time, $(k-1)$ must be placed after a_1 too because there is no element before a_1. It follows that $(k-2)$ must be placed after

Chapter A: Solutions

$(k-1)$ because otherwise there will be no element after $(k-1)$ which differs it by just 1. Repeating this process will yield that $a_n < a_1 = k$. This contradicts to the earlier conclusion. This means that a_1 must be either 1 or n or

$$S_n = 2S_{n-1} \implies \boxed{S_n = 2^{n-1}}$$

Practice 9

Find a recursion for the last practice in the previous chapter (no need to solve the recursion for now):

Let $x_i \in \{+1, -1\}$, $i = 1, 2, \cdots, 2n$. If their sum equals 0 and the following inequality holds for any positive integer k satisfying $1 \le k < 2n$:

$$x_1 + x_2 + \cdots + x_k \ge 0$$

Find the number of possible ordered sequence $\{x_1, x_2, \cdots, x_{2n}\}$.

Let the desired count be a_n.

By the definition, we must have $x_1 = 1$. Let $m \le n$ be the smallest integer such that

$$x_1 + x_2 + \cdots + x_{2m} = 0$$

Then $x_{2m} = -1$. This means that $x_2 + x_3 + \cdots + x_{2m-1} = 0$ and, for any positive integer $i \le 2m - 2$, it holds $x_2 + \cdots + x_i \ge 0$. Then, the set containing $2(m-1)$ ordered numbers $\{x_2, x_3, \cdots, x_{2m-1}\}$ satisfies the original requirements. Therefore, there are a_{m-1} such collections.

Similarly, the remaining set $\{x_{2m+1}, x_{2m+2}, \cdots, x_{2n}\}$ also satisfies the given requirements which means there are a_{n-m} such sets.

It follows that a_n satisfies

$$a_n = \sum_{m=1}^{n} a_{m-1} a_{n-m}$$

Chapter A: Solutions

This recursion can be solved using generating function which will be discussed later.

Practice 10

Find the number of ways to divide a convex n-sided polygon into $(n-2)$ triangles using non-intersecting diagonals.

(ref: 4465)

Let the result be a_n and the n vertices be $A_1, A_2, \cdots, A_n$.

If there is no diagonal drawn from A_1, then there are a_{n-1} ways to to divide the $(n-1)$-gon $A_2 A_3 \cdots A_n$ into triangles, plus one additional triangle $\triangle A_1 A_2 A_n$.

Otherwise, if there is at least one diagonal drawn from A_1, let k be the smallest index where $A_1 A_k$ is drawn. Then, $3 \leq k \leq n-1$ and there must exist $\triangle A_1 A_2 A_k$. Meanwhile, the $(k-1)$-polygon $A_2 A_3 \cdots A_k$ has a_{k-1} ways to divide into triangles and the $(n-k+2)$-polygon $A_1 A_k A_{k+1} \cdots A_n$ has a_{n-k+2} ways. Therefore, we have

$$a_n = a_{n-1} + \sum_{k=3}^{n-1} a_{k-1} a_{n-k+2} = a_{n-1} + \underbrace{\sum_{j=2}^{n-2} a_j a_{n-j+1}}_{\text{let } j=k-1}$$

where we define $a_2 = 1$ (thinking about when the case when $k = 3$ or $k = n-2$). It follows that $a_{n-1} = a_2 a_{n-1}$. Therefore, the above recursion can be written as

$$a_n = \sum_{j=2}^{n-1} a_j a_{n-j+1}$$

where $a_2 = 1$ and $a_3 = 1$. This recursion essentially the same as that in the previous practice except the index differs by 2. Based on the conclusion of the last practice in the previous chapter, the

Chapter A: Solutions

answer to this problem is

$$\boxed{\frac{1}{n-1}\binom{2(n-2)}{n-2}}$$

A.4 Chapter 4

Practice 1

Find the number of positive integer solutions to the following equation:
$$x_1 + x_2 + \cdots + x_5 = 14$$

(ref: 2526)

This is a basic pattern and the answer is
$$\binom{14-1}{5-1} = \boxed{715}$$

Practice 2

Find the number of non-negative integer solutions to the following equation:
$$x_1 + x_2 + \cdots + x_5 = 14$$

(ref: 4134)

This is a basic pattern and the answer is
$$\binom{14+5-1}{5-1} = \boxed{3060}$$

Practice 3

Find the number of integer solutions to the following equation:
$$x_1 + x_2 + \cdots + x_6 = 12$$

where $x_1, x_5 \geq 0$ and $x_2, x_3, x_4 > 0$

(ref: 4135)

Chapter A: Solutions

This problem can be transformed to a case of counting all positive integer solutions or counting all non-negative integer solutions.

Solution 1

Let $y_1 = x_1 + 1$ and $y_5 = x_5 + 1$, Then the original equation becomes

$$y_1 + x_2 + x_3 + x_4 + y_5 = 14$$

where all the variables are positive integers. This is a basic pattern and the answer is

$$\binom{14-1}{5-1} = \boxed{715}$$

Solution 2

Let $y_2 = x_2 - 1, y_3 = x_3 - 1$, and $y_4 = x_4 - 1$. Then the original equation becomes

$$x_1 + y_2 + y_3 + y_4 + x_5 = 9$$

where all variables are non-negative integers. This is a basic pattern and the answer is

$$\binom{9+(5-1)}{5-1} = \boxed{715}$$

These two answers agree with each other.

Practice 4

Find the number of non-negative integer solutions to the equation

$$2x_1 + x_2 + x_3 + \cdots + x_9 + x_{10} = 3$$

(ref: 4409)

When $x_1 \geq 2$, this equation has no no non-negative integer solution.

When $x_1 = 0$, the original equation is equivalent to

$$x_2 + x_3 + \cdots + x_9 + x_{10} = 3$$

Chapter A: Solutions

which has $\binom{3+9-1}{9-1} = 165$

When $x_1 = 1$, the original equation is equivalent to

$$x_2 + x_3 + \cdots + x_9 + x_{10} = 1$$

which has 9 non-negative integer solutions.

Therefore, the final answer is

$$165 + 9 = \boxed{174}$$

Practice 5

Pat is to select six cookies from a tray containing only chocolate chip, oatmeal, and peanut butter cookies. There are at least six of each of these three kinds of cookies on the tray. How many different assortments of six cookies can be selected?

(ref: 2824 - AMC10)

This is equivalent to counting the number of non-negative integer solutions to the following equation

$$x + y + z = 6$$

where x, y, and z represent the number of the three types of cookies, respectively. The answer as

$$\binom{6+3-1}{3-1} = \boxed{28}$$

Practice 6

Find the number of ordered quadruples of integer (a, b, c, d) satisfying $1 \leq a < b < c < d \leq 10$.

(ref: 4276)

113

Chapter A: Solutions

Let $x_1 = a$, $x_2 = b - a$, $x_3 = c - b$, $x_4 = d - c$, and $x_5 = 11 - d$. Then we have
$$x_1 + x_2 + x_3 + x_4 + x_5 = 11$$

Meanwhile, all x_i, $(i = 1, 2, 3, 4, 5)$ are positive integers. Therefore the desired answer is the number of positive integer solutions to the above equation which is
$$\binom{11-1}{5-1} = \boxed{210}$$

Practice 7

How many different outcomes are there if three dices are rolled at the same time?

(ref: 4419)

This is equivalent to the number of non-negative integer solutions to the following equation
$$x_1 + x_2 + x_3 + x_4 + x_5 + x_6 = 3$$
where x_k, $(1 \leq k \leq 6)$ indicates the number of the dies whose reading is k. Therefore, the answer is
$$\binom{3+6-1}{3} = \boxed{56}$$

Practice 8

A total of 2018 tickets, numbered 1, 2, 3, $\cdots$, 2017, 2018 are placed in an empty bag. Alfred removes ticket a from the bag. Bernice then removes ticket b from the bag. Finally, Charlie removes ticket c from the bag. They notice that $a < b < c$ and $a + b + c = 2018$. In how many ways could this happen?

(ref: 2679)

Chapter A: Solutions

This is equivalent to counting the number of ordered and distinct positive integer solutions to the equation

$$a + b + c = 2018$$

The total count, without any restriction, is

$$\binom{2018-1}{3-1} = \binom{2017}{2}$$

Because 2018 is not divisible by 3, therefore a, b, and c cannot be all equal. Hence, there are two possibilities: (1) all the numbers are distinct, and (2) two of them are the same.

The 2nd case is easy to count because the same number can only be 1, 2, $\cdots$, 1008. Consequently, the total number of possibilities where two numbers are equal is

$$1008 \cdot \binom{3}{1}$$

In the 1st case, when a, b and c are distinct, each triplet (a, b, c) can form $3! = 6$ different solutions of which only 1 is ordered. This means only $\frac{1}{6}$ of the cases in this category satisfies the condition $a < b < c$.

Hence, the final answer to the original question is

$$\frac{1}{6} \cdot \left(\binom{2017}{2} - 3 \times 1008 \right) = \boxed{338352}$$

Practice 9

Find the number of non-decrease sequences of length n and each element is a non-negative integer not exceeding d.

(ref: 4145)

The answer is $\boxed{\binom{n+d}{n}}$ or $\boxed{\binom{n+d}{d}}$.

Chapter A: Solutions

It is equivalent to finding the number of non-negative integer solutions to the following equation

$$x_0 + x_1 + \cdots + x_d = n$$

In this model, each x_i ($0 \leq i \leq d$) indicates the number of elements in this sequence whose value equals i. Because total n elements need to be selected, therefore they should sum to n. Meanwhile, given any selection (i.e. one solution to the above equation), there is only one way to form a non-decreasing sequence. Hence the number of non-decreasing sequence equals the number of non-negative integers solutions to the above equation.

Practice 10

Let $\mathbb{A} = \{a_1, a_2, \cdots, a_{100}\}$ be a set containing 100 real numbers, $\mathbb{B} = \{b_1, b_2, \cdots, b_{50}\}$ be a set containing 50 real numbers, and $\mathcal{F}$ be a mapping from $\mathbb{A}$ to $\mathbb{B}$. Find the number of possible $\mathcal{F}$ if $\mathcal{F}(a_1) \leq \mathcal{F}(a_2) \leq \cdots \mathcal{F}(a_1)$, and for every $b_i \in \mathbb{B}$, there exists an element $a_i \in \mathbb{A}$ such that the $\mathcal{F}(a_i) = b_i$.

(ref: 4361 - China)

Without loss of generality, let's assume $b_1 < b_2 < \cdots < b_{50}$. Then the answer is the positive integer solutions to the following equation

$$x_1 + x_2 + \cdots + x_{50} = 100$$

where x_i denotes the number of $a_j \in \mathbb{A}$ such that $\mathcal{F}(a_j) = b_i$. Hence the answer is

$$\binom{100-1}{50-1} = \boxed{\binom{99}{49}}$$

Chapter A: Solutions

Practice 11

How many ordered integers (x_1, x_2, x_3, x_4) are there such that $0 < x_1 \leq x_2 \leq x_3 \leq x_4 < 7$?

(ref: 4411)

The answer is the same as the number of ways to select four integers from $\mathbb{S} = \{1, 2, 3, 4, 5, 6\}$ when duplication is permitted. This is because for every such selection, there is a unique way to arrange these four numbers in and increasing order. On the other hand, all qualifying x_1, x_2, x_3 and x_4 must be selected from $\mathbb{S}$.

Meanwhile, the number of ways to select 4 elements from $\mathbb{S}$, duplication permitted, is equivalent to the number of non-negative integer solutions to the following equation where k_i ($i = 1, 2, \cdots, 6$) indicates the number of i selected:

$$k_1 + k_2 + k_3 + k_4 + k_5 + k_6 = 4$$

Therefore the answer is

$$\binom{4+6-1}{4} = \boxed{126}$$

Practice 12

Define an ordered quadruple of integers (a, b, c, d) as interesting if $1 \leq a < b < c < d \leq 10$, and $a + d > b + c$. How many interesting ordered quadruples are there?

(ref: 260 - AIME)

Let $x_1 = a$, $x_2 = b - a$, $x_3 = c - b$, $x_4 = d - c$, and $x_5 = 11 - d$. Then, it holds that

$$x_1 + x_2 + x_3 + x_4 + x_5 = 11$$

Chapter A: Solutions

There are $\binom{11-1}{5-1} = 210$ positive integer solutions to this equation. However, following possibilities need to be excluded where

$$a + d > b + c \implies d - c > b - a \implies x_4 > x_2$$

By symmetry, there are as many quadruples where $x_4 > x_2$ as those where $x_2 > x_4$. Meanwhile, those cases where $x_2 = x_4$ can be manually counted:

- $x_2 = x_4 = 1 \implies x_1 + x_3 + x_5 = 9 \implies \binom{9-1}{3-1} = 28$
- $x_2 = x_4 = 2 \implies x_1 + x_3 + x_5 = 7 \implies \binom{7-1}{3-1} = 15$
- $x_2 = x_4 = 3 \implies x_1 + x_3 + x_5 = 5 \implies \binom{5-1}{3-1} = 6$
- $x_2 = x_4 = 4 \implies x_1 + x_3 + x_5 = 3 \implies 1$

Therefore, the number of interesting quadruples is

$$(210 - 28 - 15 - 6 - 1) \div 2 = \boxed{80}$$

Practice 13

How many ways are there to arrange 8 girls and 25 boys to sit around a table so that there are at least 2 boys between any pair of girls? If a sitting plan can be simply rotated to match another one, these two are treated as the same.

(ref: 4366 - China)

There are 8 intervals between these girls. Let the number of boys in each of these intervals be x_i, $i = 1, 2, \cdots, 8$, then

$$x_1 + x_2 + \cdots + x_8 = 25$$

conditioned on $x_i \geq 2$. Let $y_i = x_i - 2 \geq 0$, then

$$y_1 + y_2 + \cdots + y_8 = 9$$

Chapter A: Solutions

There are $\binom{16}{7}$ non-negative solutions to the above equation.

Meanwhile, there are $(8-1)! = 7!$ ways to arrange 8 girls in a round manner and 25! ways to arrange these boys. Hence, the final answer is

$$\binom{16}{7}(7!)(25!) = \boxed{\frac{16! \cdot 25!}{9!}}$$

Practice 14

Six men and some number of women stand in a line in random order. Let p be the probability that a group of at least four men stand together in the line, given that every man stands next to at least one other man. Find the least number of women in the line such that p does not exceed 1 percent.

(ref: 251 - AIME)

Let n be the number of women in the group.

By the given condition, there is no isolated man in the line. Therefore, five different cases are possible depending on how these six men are standing together in groups: $(2)(2)(2)$, $(2)(4)$, $(3)(3)$, $(4)(2)$, and (6). Let x_i, $i = 1, 2, \cdots$, be the number of woman standing before the first man, between two adjacent men's groups, and after the last men. Then we have 5 possible equations:

$$(2)(2)(2) \implies x_1 + x_2 + x_3 + x_4 = n$$
$$(2)(4) \implies x_1 + x_2 + x_3 = n$$
$$(3)(3) \implies x_1 + x_2 + x_3 = n$$
$$(4)(2) \implies x_1 + x_2 + x_3 = n$$
$$(6) \implies x_1 + x_2 = n$$

In these equations, the first x_1 and the last x_i can be 0 or positive (a value of 0 means no girl standing at that end), all the middle x_i must be positive integers. These are all basic pattern of counting integer solutions. Their solutions are $\binom{n+1}{3}$, $\binom{n+1}{2}$, $\binom{n+1}{2}$, $\binom{n+1}{2}$.

Chapter A: Solutions

and $\binom{n+1}{1}$, respectively.

Therefore, the desired answer is the smallest integer n satisfying:

$$\frac{2 \times \binom{n+1}{2} + \binom{n+1}{1}}{\binom{n+1}{3} + 3 \times \binom{n+1}{2} + \binom{n+1}{1}} \leq \frac{1}{100}$$

This relation can be simplified to

$$n(n - 592) \geq 594$$

It follows that $n \geq 593$ in order to make the left side be positive. Then it can be determined that $n = \boxed{594}$ is the desired answer.

Practice 15

How many different ways to write a positive integer n as a sum of m different positive integers? Different sequences are treated as distinct.

(ref: 4413)

This is a special case of integer partition. Under these two conditions (fixed number of partitions and distinct sequences), this problem can be modeled as counting the positive integer solutions to

$$x_1 + x_2 + \cdots + x_m = n$$

whose answer is $\boxed{\binom{n-1}{m-1}}$.

A.5 Chapter 5

Practice 1

Simplify the expression

$$S = \binom{2020}{0}^2 + \binom{2020}{1}^2 + \cdots + \binom{2020}{2020}^2$$

(ref: 2721)

Applying Vandermonde's identity, *(5.3)* on *page 44*, yields

$$S = \binom{2020}{0}\binom{2020}{2020} + \binom{2020}{1}\binom{2020}{2019} + \cdots + \binom{2020}{0}\binom{2020}{2020} = \binom{2020 \times 2}{2020} = \boxed{\binom{4040}{2020}}$$

Practice 2

Given

$$P(x) = (1 + x + x^2)^{100} = a_0 + a_1 x + \cdots + a_{200} x^{200}$$

Compute the following sums:

- $S_1 = a_0 + a_1 + a_2 + a_3 + \cdots + a_{200}$
- $S_2 = a_0 + a_2 + a_4 + a_6 + \cdots + a_{200}.$

(ref: 3978)

Setting $x = 1$ gives the result for S_1:

$$S_1 = a_0 + a_1 + a_2 + \cdots + a_{200} = \boxed{3^{100}}$$

Next, setting $x = -1$ yields

$$S_{-1} = a_0 - a_1 + a_2 + \cdots + a_{200} = 1^{100} = 1$$

Chapter A: Solutions

Now, we have

$$S_2 = \frac{1}{2} \cdot (S_1 + S_{-1}) = \boxed{\frac{3^{100} + 1}{2}}$$

Practice 3

For some particular value of N, when $(a+b+c+d+1)^N$ is expanded and like terms are combined, the resulting expression contains exactly 1001 terms that include all four variables $a, b, c,$ and d, each to some positive power. What is N?

(ref: 2884 - AMC10)

After expansion, all the merged terms must be in the form of

$$a^u b^v c^w d^x 1^y$$

where integers $u + v + w + x + y = N$.

To satisfy the requirement, u, v, w, x must be positive and y can be either 0 or positive. Let $z = y + 1$, then the number of terms equals to the positive integer solutions to the following equation:

$$u + v + w + x + z = N + 1$$

This is a basic pattern and the answer is $\binom{N}{4}$. Hence,

$$\binom{N}{4} = 1001 \implies N = \boxed{14}$$

Practice 4

For any positive integer n and prime p, it must hold that

$$(1+n)^p \equiv 1 + n^p \pmod{p}$$

Chapter A: Solutions

The conclusion of *Example 5.5.1* on *page 48* states that

$$\binom{p}{k} \equiv 0 \pmod{p}$$

for all k where $0 < k < p$. Therefore

$$(1+n)^p = 1 + \binom{n}{1}n + \cdots + \binom{n}{n-1}n^{p-1} + n^p \equiv 1 + n^p \pmod{p}$$

Practice 5

Evaluate the value of

$$\sum_{m=0}^{2009} \sum_{n=0}^{m} \binom{2009}{m}\binom{m}{n}$$

(ref: 4323 - PUMaC)

Firstly, this expression can be rewritten as

$$\sum_{m=0}^{2009} \sum_{n=0}^{m} \binom{2009}{m}\binom{m}{n} = \sum_{m=0}^{2009} \binom{2009}{m} \sum_{n=0}^{m} \binom{m}{n}$$

The inner part is the sum of coefficients. Applying *Equation 5.8* on *page 50* yields

$$\sum_{m=0}^{2009} \binom{2009}{m} \sum_{n=0}^{m} \binom{m}{n} = \sum_{m=0}^{2009} \binom{2009}{m} 2^m$$

Now this expression is the sum of all the coefficients of $(x+2)^{2009}$ which can be obtained by setting $x = 1$. This means that the final answer is

$$(1+2)^{2009} = \boxed{3^{2009}}$$

Chapter A: Solutions

Practice 6

Let $\mathbb{S}$ be a set of integers, $\max(\mathbb{S})$ be the largest element in $\mathbb{S}$, and $|\mathbb{S}|$ be the number of elements in $\mathbb{S}$. Find the number of non-empty set $\mathbb{S} \in \{1, 2, \cdots, 10\}$ satisfying $\max(\mathbb{S}) \leq |\mathbb{S}| + 2$.

(ref: 4327 - PUMaC)

Let $k = |\mathbb{S}|$. When $k \geq 9$, the requirements will be automatically satisfied. For each value in $1 \leq k \leq 8$, the elements in corresponding $\mathbb{S}$ can only be selected from $1, 2, \cdots, k+2$. There are $\binom{k+2}{k}$ different selections. It follows that the desired answer is

$$\binom{10}{10} + \binom{10}{9} + \sum_{k=1}^{8}\binom{k+2}{k} = 11 + \sum_{k=1}^{8}\binom{k+2}{2} = 11 + \left(\binom{11}{3} - \binom{2}{2}\right) = \boxed{175}$$

The second last step utilizes the hockey stick identity.

Practice 7

Let m and n be positive integers. Show that

$$\frac{(m+n)!}{(m+n)^{m+n}} < \frac{m!\, n!}{m^m\, n^n}$$

(ref: 4328 - Putnam)

This inequality is equivalent to

$$(m+n)^{m+n} > \frac{(m+n)!}{m!\,n!} m^m n^n = \binom{m+n}{m,n} m^m n^n$$

which indeed holds because the expansion of the left side contains the right side plus some other positive terms.

Chapter A: Solutions

Practice 8

Let n be an even integer. Find the number of ways to select four distinct integers a, b, c, d between 1 and n, inclusive, satisfying $a + c = b + d$. Order of these four numbers does not matter.

(ref: 4417)

Without loss of generality, assuming $a < c$ and $b < d$. Let $s = a + c = b + d$. Then it must hold that $5 \leq s \leq (2n - 3)$. The middle value of s is $(n + 1)$.

- When $s < (n + 1)$, both a and b must be chosen from 1 and $\left\lfloor \frac{s-1}{2} \right\rfloor$, inclusive.

- When $s > (n + 1)$, both c and d must be chose from $\left\lfloor \frac{s-1}{2} \right\rfloor$ to n, inclusive.

- When $s = (n + 1)$, choosing a and b first or c and d first will result the same.

When either a and b, or c and d are chosen, the other two values will be determined too because their respective sums are s. So, when s loops through $5, 6, \cdots, (2n - 4), (2n - 3)$, the number of possible selections will be

$$\binom{2}{2}, \binom{2}{2}, \cdots, \binom{\lfloor \frac{n}{2} \rfloor - 1}{2}, \binom{\lfloor \frac{n}{2} \rfloor - 1}{2}, \binom{\lfloor \frac{n}{2} \rfloor}{2} \binom{\lfloor \frac{n}{2} \rfloor - 1}{2}, \binom{\lfloor \frac{n}{2} \rfloor - 1}{2}, \cdots \binom{2}{2}, \binom{2}{2}$$

It follows that their sum, i.e. the final result, is

$$4\left(\binom{2}{2} + \binom{3}{2} + \cdots + \binom{\lfloor \frac{n}{2} \rfloor - 1}{2}\right) + \binom{\lfloor \frac{n}{2} \rfloor}{2}$$

$$= 4\binom{\lfloor \frac{n}{2} \rfloor}{3} + \binom{\lfloor \frac{n}{2} \rfloor}{2} \quad \text{(by hockey stick)}$$

$$= 4 \cdot \frac{\frac{n}{2}\left(\frac{n}{2} - 1\right)\left(\frac{n}{2} - 2\right)}{3!} + \frac{\frac{n}{2}\left(\frac{n}{2} - 1\right)}{2!}$$

$$= \boxed{\frac{n(n-2)(2n-5)}{24}}$$

Chapter A: Solutions

Practice 9

Find the sum of all n such that

$$\binom{n}{0} - \binom{n}{1} + \binom{n}{2} - \binom{n}{3} + \cdots + \binom{n}{2018} = 0$$

(ref: 4324)

This is a 2018th degree polynomial which has 2018 complex roots. Because all integers n satisfying $1 \leq n \leq 2018$ meet the requirement, therefore the desired answer is

$$1 + 2 + \cdots + 2018 = \boxed{2037171}$$

Practice 10

How many 3×3 matrices of non-negative integers are there such that the sum of every row and every column equals n?

(ref: 4425)

Let the matrix be

$$\begin{bmatrix} a_1 & b_1 & c_1 \\ a_2 & b_2 & c_2 \\ a_3 & b_3 & c_3 \end{bmatrix}$$

First, let's consider the top two rows. There are $\binom{n+2}{2}$ solutions each to the following two equations

$$a_1 + b_1 + c_1 = n \quad \text{and} \quad a_2 + b_2 + c_2 = n$$

Because the sum of each column is n, therefore fixing the top two rows will determine the third row. However, we need to exclude those cases where the third row contains negative elements.

Without loss of generality, let's assume $a_3 < 0$. This means that

$$a_1 + a_2 = n - a_3 \geq n_1$$

126

Chapter A: Solutions

Then
$$b_1 + c_1 + b_2 + c_2 = 2n - (a_1 + a_2) \leq n - 1$$

which implies both b_3 and c_3 will have to be positive. So there is at most one negative element in the third row. The number of such cases equal to the non-negative integer solutions to

$$b_1 + c_1 + b_2 + c_2 = 2n - (a_1 + a_2) = n + a_3$$

where $-n \leq a_3 \leq -1$ which equals

$$\sum_{a_3=-n}^{-1} \binom{n + a_3 + 3}{3} = \binom{n+3}{4}$$

There are three cases of which element in the third row can be negative. Therefore, the final answer is

$$\boxed{\binom{n+2}{2}^2 - 3\binom{n+3}{4}}$$

Practice 11

Show that for any positive integer n, the value of $\frac{(n^2)!}{(n!)^{n+1}}$ is an integer.

(ref: 4363)

Let's consider the following counting problem: how many different ways to distribute n^2 distinguishable balls equally among n indistinguishable boxes.

This problem is similar to the "equally distributing 6 distinct books among 3 piles" problem in the first book. Therefore, the answer is

$$\frac{1}{n!} \cdot \frac{(n^2)!}{n! \cdot n! \cdots n!} = \frac{(n^2)!}{(n!)^{n+1}}$$

Therefore, this value must be an integer.

Chapter A: Solutions

A.6 Chapter 6

Practice 1

How many different 5-digit numbers can be formed using 1, 2, 3, and 4 that satisfy the following conditions:

- the digit 1 must appear either 2 or 3 times,
- the digit 2 must appear even times,
- the digit 3 must appear odd times, and
- the digit 4 has no restriction

(ref: 2536)

The result is the coefficient of x^5 in the expanded form of the following polynomial which is $\boxed{4}$:

$$(x^2 + x^3)(x^0 + x^2 + x^4)(x^1 + x^3 + x^5)(x^0 + x^1 + x^2 + x^3 + x^4 + x^5)$$

The four brackets are corresponding to the four digits, respectively. Within each bracket, the exponent of each term represents the number of times this digit appears. Let this polynomial be $f(x)$, then

$$f(x) = x^3(1+x)(1+x^2+x^4)(1+x^2+x^4)(1+x+x^2+x^3+x^4+x^5)$$

The x^5 term in $f(x)$ must be the x^2 in

$$g(x) = (1+x)(1+x^2)(1+x^2)(1+x+x^2)$$

Here, terms with exponents higher than 2 in each bracket have already discarded. Now if 1 is chosen from the first bracket, then among the other three brackets one of the must contribute an x^2 term and the other two must contribute the constant 1 in order to obtain an x^2 term. The sum of coefficients in these cases is 3. When the first bracket contributes the x term, then the second and forth brackets must also contribute 1 and third bracket must contribute the x term. The coefficient in this case is 1. Adding them together gives the final result as $\boxed{4}$.

Note that, given the limited number of possibilities, this problem can also be solved using casework.

Practice 2

Find the number of integer solutions to the equation $a+b+c = 6$ where $-1 \leq a < 2$ and $1 \leq b,\ c \leq 4$.

(ref: 4360)

The answer is the coefficient of the x^6 term in

$$f(x) = (x^{-1} + 1 + x + x^2)(x + x^2 + x^3 + x^4)^2$$
$$= x\left(1 + x + x^2 + x^3\right)^3$$

which is the same as the coefficient of the x^5 term in

$$g(x) = \left(1 + x + x^2 + x^3\right)^3$$
$$= \frac{1}{(1-x)^3} \cdot (1-x^4)^3$$
$$= \left(\sum_{k=0}^{\infty} \binom{-3}{k}(-1)^k x^k\right)\left(1 - 3x^4 + 3x^8 - x^{12}\right)$$

Therefore, the answer is

$$\binom{-3}{5}(-1)^5 - 3 \cdot \binom{-3}{1}(-1)^1 = \boxed{12}$$

This result can be verified by listing all the qualified solutions:

$(-1, 3, 4),\ (-1, 4, 3),\ (0, 2, 4),\ (0, 3, 3),\ (0, 4, 2), (1, 1, 4),$
$(1, 2, 3),\ (1, 3, 2),\ (1, 4, 1),\ (2, 1, 3),\ (2, 2, 2),\ (2, 3, 1)$

Chapter A: Solutions

Practice 3

How many different ways are there to make a payment of n dollars using any number of $1 and $2 bills?

(ref: 4329)

The answer is the coefficient of the x^n term in the expansion of

$$f(x) = (1 + x + x^2 + \cdots)(1 + x^2 + x^4 + \cdots)$$
$$= \frac{1}{1-x} \cdot \frac{1}{1-x^2}$$
$$= \frac{1}{(1-x)^2} \cdot \frac{1}{1+x}$$
$$= \frac{1}{2}\left(\frac{1}{(1-x)^2} + \frac{1}{1-x^2}\right)$$
$$= \frac{1}{2}\left((1 + 2x + 3x^2 + 4x^3 + \cdots)\right.$$
$$\left. + (1 + x^2 + x^4 + x^6 + \cdots)\right) \qquad (\because 6.3,\ 6.1)$$
$$= 1 + x + 2x^2 + 2x^3 + 3x^4 + \cdots$$
$$= \sum_{n=0}^{\infty} \left(\left\lfloor \frac{n}{2} \right\rfloor + 1\right) x^n$$

Therefore, the answer is $\left(\left\lfloor \dfrac{n}{2} \right\rfloor + 1\right)$ where function $\lfloor x \rfloor$ returns the largest integer not exceeding the given real number x.

Practice 4

How many 4-digit integers are there whose sum of all digits equals 12?

(ref: 4330)

This is equivalent to find the coefficient of the x^{12} term in the following polynomial:

$$f(x) = (x + x^2 + \cdots + x^9)(1 + x + x^2 + \cdots + x^9)^3$$

because the thousands digit can not be 0 and all the remaining 3 digits can be any of 0 to 9.

$$\begin{aligned}
f(x) &= (x + x^2 + \cdots + x^9)(1 + x + x^2 + \cdots + x^9)^3 \\
&= \frac{x - x^{10}}{1 - x}\left(\frac{1 - x^{10}}{1 - x}\right)^3 \\
&= \frac{(x - x^{10})(1 - x^{10})^3}{(1 - x)^4} \\
&= x(x^{39} - x^{30} - 3x^{29} + 3x^{20} + 3x^{19} - 3x^{10} - x^9 + 1)\sum_{k=0}^{\infty}\binom{3+k}{3}x^k
\end{aligned}$$

Therefore, the coefficient of the term x^{12} equals

$$-3 \cdot \binom{3+1}{3} - \binom{3+2}{3} + \binom{3+11}{3} = \boxed{342}$$

Practice 5

There are 30 identical souvenirs to be distributed among 50 students. Each student may receive more than one souvenir. How many different distribution plans are there?

(ref: 4359)

The result is the coefficient of the term x^{30} in the expanded form of the following polynomial

$$(1 + x + x^2 + \cdots)^{50} = \left(\frac{1}{1 - x}\right)^{50} = (1 - x)^{-50}$$

The generalized binomial theorem gives the answer as

$$\binom{-50}{30}(-1)^{30} = \binom{50 + 30 - 1}{30} = \boxed{\binom{79}{30}}$$

Chapter A: Solutions

Practice 6

A deck of poker has three different colors each of which contains 10 cards numbered from 1 to 10, respectively. In addition, there are two jokers both of which are numbered as 0. A card with number k is valued as 2^k points. How many different ways are there to draw several cards from this deck so that their total value equals 2004?

(ref: 4333 - China)

Let a_n be the number of combinations of cards such that their total value is n. Then we have

$$\sum_{k=0}^{\infty} a_k x^k = \left(1+x^{2^0}\right)^2 \left(1+x^{2^1}\right)^3 \left(1+x^{2^2}\right)^3 \cdots \left(1+x^{2^{10}}\right)^3$$

$$= \frac{1}{1+x}\left(\left(1+x^{2^0}\right)\left(1+x^{2^1}\right)\left(1+x^{2^2}\right)\cdots\left(1+x^{2^{10}}\right)\right)^3$$

$$= \frac{1}{1+x}\frac{1}{(1-x)^3}\left(1-x^{2^{11}}\right)^3$$

Because $2004 < 2^{11}$, therefore a_{2004} must equal the coefficient of the x^{2004} in the expanded form of

$$\frac{1}{(1+x)(1-x)^3}$$
$$= \frac{1}{1-x^2} \cdot \frac{1}{(1-x)^2}$$
$$= (1+x^2+x^4+\cdots)\left(1+\binom{1+1}{1}x+\binom{1+2}{1}x^2+\binom{1+3}{1}x^3+\cdots\right)$$

Therefore

$$a_{2004} = \binom{1+2004}{1} + \binom{1+2002}{1} + \cdots + \binom{1+2}{1} + \binom{1+0}{1}$$
$$= 2005 + 2003 + \cdots + 3 + 1$$
$$= 1003^2$$
$$= \boxed{1006009}$$

Chapter A: Solutions

Practice 7

Let n be a positive integer. Find the number a_n of polynomials $f(x)$ with coefficients in $\{0, 1, 2, 3\}$ such that $f(2) = n$.

(ref: 4335)

Let
$$f(x) = c_0 + c_1 x + c_2 x^2 + \cdots + c_k x^k$$
and $c_i \in \{0, 1, 2, 3\}$ where $i = 0, 1, 2, \cdots, k$. Then, $f(2) = n$ is equivalent to
$$c_0 + 2c_1 + 2^2 c_2 + \cdots + 2^k c_k = n$$
Note that every $2^i c_i$ can only take a value of 0, 2^i, $2 \cdot 2^i$, or $3 \cdot 2^i$. Thus, a_n is the coefficient of the term t^n in the following polynomial
$$P(x) = \prod_{i=0}^{\infty} \left(1 + t^{2^i} + t^{2 \cdot 2^i} + t^{3 \cdot 2^i}\right)$$
Applying the sum of geometric sequence formula gives
$$P(x) = \frac{1-t^4}{1-t} \frac{1-t^8}{1-t^2} \frac{1-t^{16}}{1-t^4} \frac{1-t^{32}}{1-t^8} \cdots$$
$$= \frac{1}{1-t} \frac{1}{1-t^2}$$
$$= \left(1 + t + t^2 + t^3 + \cdots\right)\left(1 + t^2 + t^4 + t^6 + \cdots\right)$$
$$= 1 + t + 2t^2 + 2t^3 + 3t^4 + 3t^5 + 4t^6 + \cdots$$

Therefore, the final answer is
$$\boxed{a_n = \left\lfloor \frac{n}{2} \right\rfloor + 1}$$

Chapter A: Solutions

Practice 8

Let $a_0, a_1, a_2, \cdots$ be an increasing sequence of non-negative integers such that every non-negative integer can be expressed uniquely in the form of $(a_i + 2a_j + 4a_k)$ where i, j, and k are not necessarily distinct. Determine a_{1998}.

(ref: 4336 - IMO)

Let
$$f(x) = \sum_{i=0}^{\infty} x^{a_i}$$

Then the given condition implies
$$f(x)f(x^2)f(x^4) = \sum_{n=0}^{\infty} x^n = \frac{1}{1-x}$$

Replacing x with x^2 gives
$$f(x^2)f(x^4)f(x^8) = \frac{1}{1-x^2}$$

These two relationships above imply
$$f(x) = (1+x)f(x^8)$$

Repeating this process recursively will give
$$f(x) = (1+x)\left(1+x^8\right)\left(1+x^{8^2}\right)\left(1+x^{8^3}\right)\cdots$$

$$\therefore \sum_{i=0}^{\infty} x^{a_i} = (1+x)\left(1+x^8\right)\left(1+x^{8^2}\right)\left(1+x^{8^3}\right)\cdots$$

Expanding the right side shows that a_i are those non-negative integers whose base 8 representation has only digit 0 or 1.

As $1998 = 2 + 2^2 + 2^3 + 2^6 + 2^7 + 2^8 + 2^9 + 2^{10}$, therefore

$$a_{1998} = \boxed{8 + 8^2 + 8^3 + 8^6 + 8^7 + 8^8 + 8^9 + 8^{10}}$$

Practice 9

Show that

$$\sqrt{1+x} = 1 + \sum_{n=1}^{\infty} \frac{(-1)^{n-1}}{n \cdot 2^{2n-1}} \binom{2n-2}{n-1} x^n$$

(ref: 4314)

Applying the generalized binomial expansion yields

$$(1+x)^{\frac{1}{2}} = \sum_{n=0}^{\infty} \binom{\frac{1}{2}}{n} x^n = 1 + \sum_{n=1}^{\infty} \binom{\frac{1}{2}}{n} x^n$$

Now the coefficient can be transformed as

$$\begin{aligned}
\binom{\frac{1}{2}}{n} &= \frac{\left(\frac{1}{2}\right)\left(\frac{1}{2}-1\right)\left(\frac{1}{2}-2\right)\cdots\left(\frac{1}{2}-(n-1)\right)}{n!} \\
&= \frac{(-1)^{n-1}}{2^n} \cdot \frac{1 \cdot 3 \cdot 5 \cdots (2n-3)}{n!} \\
&= \frac{(-1)^{n-1}}{2^n \cdot n!} \cdot \frac{(2n-2)!}{2 \cdot 4 \cdot 6 \cdots (2n-2)} \\
&= \frac{(-1)^{n-1}}{2^n \cdot n!} \cdot \frac{(2n-2)!}{(2 \cdot 1)(2 \cdot 2)(2 \cdot 3) \cdots (2 \cdot (n-1))} \\
&= \frac{(-1)^{n-1}}{2^n \cdot n!} \cdot \frac{(2n-2)!}{2^{n-1} \cdot (n-1)!} \\
&= \frac{(-1)^{n-1}}{2^{2n-1} \cdot n} \cdot \frac{(2n-2)!}{(n-1)! \cdot (n-1)!} \\
&= \frac{(-1)^{n-1}}{2^{2n-1} \cdot n} \binom{2n-2}{n-1}
\end{aligned}$$

Setting this result back to the binomial expansion leads to the conclusion immediately.

Chapter A: Solutions

Practice 10

Complete the last two practices in *Chapter 3* which is to solve the recursion

$$a_n = \sum_{k=0}^{n-1} a_k a_{n-k-1} = a_0 a_{n-1} + a_1 a_{n-2} + \cdots + a_{n-1} a_0$$

where $a_0 = a_1 = 1$.

(ref: 4511)

Let
$$f(x) = a_0 x + a_1 x^2 + a_2 x^3 + \cdots + a_n x^{n+1} + \cdots$$

Then the desired result, a_n, is the coefficient of the term x^{n+1}.

$$f^2(x) = a_0^2 x^2 + (a_0 a_1 + a_1 a_0) x^3 + \cdots + \sum_{k=0}^{n-1} a_k a_{n-k-1} x^n + \cdots$$
$$= a_1 x^2 + a_2 x^3 + \cdots + a_n x^{n+1} + \cdots \quad (\because a_0 = a_1 = 1)$$
$$= f(x) - x$$

Solving this equation and also note that $f(0) = 0$ yields

$$f(x) = \frac{1 - \sqrt{1 - 4x}}{2}$$

The conclusion from the previous practice states that

$$\sqrt{1+x} = 1 + \sum_{n=1}^{\infty} \frac{(-1)^{n-1}}{n \cdot 2^{2n-1}} \binom{2n-2}{n-1} x^n$$

Replacing x with $-4x$ gives

$$\sqrt{1 - 4x} = 1 - \sum_{n=1}^{\infty} \frac{2}{n} \binom{2n-2}{n-1} x^n$$

Therefore,

$$f(x) = \sum_{n=1}^{\infty} \frac{1}{n} \binom{2(n-1)}{n-1} x^n$$

By the definition of $f(x)$, a_n is the coefficient of the term x^{n+1} which means that

$$a_n = \boxed{\frac{1}{n+1}\binom{2n}{n}}$$

which is a Catalan number.

Chapter A: Solutions

www.ingramcontent.com/pod-product-compliance
Lightning Source LLC
Chambersburg PA
CBHW030650220526
45463CB00005B/1714